Three-dimensional Separated Flow Topology

FOCUS SERIES IN FLUID MECHANICS

Three-dimensional Separated Flow Topology

Critical Points, Separation Lines and Vortical Structures

Jean Délery

ISTE

WILEY

First published 2013 in Great Britain and the United States by ISTE Ltd and John Wiley & Sons, Inc.

ISTE Ltd
27-37 St George's Road
London SW19 4EU
UK

www.iste.co.uk

John Wiley & Sons, Inc.
111 River Street
Hoboken, NJ 07030
USA

www.wiley.com

Library of Congress Control Number: 2012955534

British Library Cataloguing-in-Publication Data
A CIP record for this book is available from the British Library
ISSN: 2051-2481 (Print)
ISSN: 2051-249X (Online)
ISBN: 978-1-84821-450-7

MIX
Paper from responsible sources
FSC® C013604
www.fsc.org

Printed and bound in Great Britain by CPI Group (UK) Ltd., Croydon, Surrey CR0 4YY

Table of Contents

Introduction

The main goal of this book is primarily didactic. It does not present a complete analysis on separated three-dimensional flows, but some basic concepts to be used to arrive at a rational description of three-dimensional flows containing what are called separations. This explains the detail of some chapters, those devoted to the critical point theory in particular, and the absence of references in the text. However, a bibliography giving the basic publications is provided.

The passage from the familiar two-dimensional to the mysterious three-dimensional requires a complete reconsideration of apparently obvious concepts (separation and reattachment points, separated bubbles, recirculation zone, limiting streamline, etc.) which are inappropriate and even dangerous to use in three-dimensional flows. Such a situation entails a risk of misunderstanding, hence leading to sterile actions to interpret facts from inadequate concepts.

Robert Legendre recognized this fact in the early 1950s when he studied the flow past delta wings and other three-dimensional bodies. He introduced a precise definition of the basic concepts on which we have to reason to describe such a flow; then he applied to these concepts rigorous results inspired from the works of the mathematician Henri

Poincaré. Hence, the introduction of notions such as skin friction lines, critical points, separation (or re-attachment) lines, separation (or re-attachment) surfaces and topological rules that allowed a consistent description of flow fields. The work of Legendre on three-dimensional separation was in great part inspired and confirmed by the visualizations performed at the same time by Henri Werlé in his water tunnels at Onera.

The present theory of three-dimensional separation is not a predictive theory in that it does not predict separation on an obstacle in any given condition. For this, the Navier–Stokes equations must be solved. It is a descriptive theory reasoning on the properties of a given vector field (skin friction and velocity) provided by experimentation or calculation. It is a tool to give a rational description and interpretation of an observation made on a model in a wind tunnel or on the screen of a workstation, displaying computed results. Even with its limitations, the presented theory is of great help in analyzing complex flows, if one considers the numerous mistakes made when speaking about three-dimensional separated flows: at separation, the skin friction vanishes, the separation is open or closed, or worst it is of the bubble type. As will be seen, on a three-dimensional vehicle, whose flow is separated, the skin friction is nearly always non-zero, most separated zones are open and the bubble notion applies only to very singular mathematical situations, hence highly improbable in reality.

It is not possible to discuss separation in three-dimensional flows without introducing the critical point theory. It is to the great credit of Robert Legendre to have realized that this mathematical theory is the only rational tool that allows the understanding of the organization – or topology – of three-dimensional separated flows. The critical point theory comes from the works of Poincaré on the singular points of the system of differential equations. In general, it applies to a space of n dimensions and leads to major conclusions and results on the behavior of dynamical systems, including the concepts of strange attractors, the notions of ill- and well-posed problems, unpredictability, and ultimately chaos theory. In what follows, we restrict ourselves to a very narrow aspect of this theory, which will be sufficient for our purpose and which, in fact, was only considered by Legendre. Critical points are found in many disciplines where they can have a different denomination and physical meaning. They are present in electromagnetism, in the theory of dynamics system, in the stability theories, etc.

The first articles published in 1952 by Robert Legendre had, in France, a limited echo mainly because of their extreme conciseness and their apparently too mathematical approach. However, it seemed that these notions were too abstract – or fundamental, not to say academic! – to be applied to real configurations, engineers being then familiar with two-dimensional analyses. However, Legendre's publications were not ignored abroad where the critical point theory was revisited by Lighthill in 1963 to depict the separation of a three-dimensional boundary layer, which, in fact, was a non-justified restriction. The critical point theory applied to three-dimensional separation led to a series of publications: in the United States where Murray Tobak published a very interesting paper in 1982, and in the United Kingdom and in Germany. We also have to cite the essential contribution of Henri Werlé whose water tunnel visualizations have constituted the physical basis, and the first motivation, of Legendre's publications.

The considerations that follow apply to a non-fluctuating flow, a laminar flow – rarely achieved in aeronautical applications – as well as to time-averaged flow introduced to describe and compute turbulent flows. In this case, as far as the organization of the flow is concerned, there are no major differences between laminar and turbulent situations (the governing equations are almost the same) except for the characteristic scales. In addition, we consider the flow as steady, although the theory can be applied to the instantaneous field of a time-dependent flow.

In many circumstances, three-dimensional separation is such a catastrophic and overwhelming phenomenon that it is nearly independent of the Reynolds number, which is, in fact, the correct scaling parameter in boundary-layer-like situations where viscous effects are confined within thin layers. The Mach number also is not a determining parameter for largely separated flows, the accompanying shock system being in reality an epiphenomenon. Thus, if we exclude exceptional circumstances where the Reynolds number is extremely low, the organization of the flow over a delta wing, downstream of the base of a missile, in front of a blunt obstacle, or past an automobile, is basically the same as in a small-size water tunnel at low velocity of a few centimeters per second. To pretend that in most cases there is no Reynolds number effect would thus be an exaggeration, but the flow physics does not depend critically on this parameter. If we exclude laminar to turbulent transition, the Reynolds number has its most direct influence on the location of the boundary-layer separation. In situations where this separation is imposed by a local singularity, like a sharp edge, the Reynolds number plays a nearly negligible role.

In Chapter 1, the critical point theory is presented, showing the origin of the nodes, saddle points and foci observed in surface flow visualizations. In Chapter 2, we introduce the basic notions of separation and attachment lines, separation and attachment surfaces, and vortical structures. This allows the rational definition in Chapter 3, of a separated flow and the description of the associated field with the formation of horseshoe and tornado-like vortices. Then, the elements and grammar rules thus established are used to describe the organization, or topology, of typical separated flows. Chapter 4 analyzes the separation over slender bodies and wings (delta and classical wings), showing how the wing vortex-wake forms. Chapter 5 deals with separation induced by an obstacle and on a blunt body. This includes the description of the flow induced by protuberances, past an afterbody and an automobile. In Chapter 6, the so-called two-dimensional (planar or axisymmetric) case is reconsidered in the light of the concepts introduced by the three-dimensional theory.

1

Skin Friction Lines Pattern and Critical Points

This chapter introduces the basic physical features underlying the topological description of three-dimensional flows, such as the skin friction lines and the surface flow pattern revealed by surface flow visualizations. It is shown that the construction of the skin friction lines reveals the existence of critical points where the skin friction vector vanishes. The mathematical properties of these points are examined and their physical meaning emphasized.

1.1. Basic properties of the three-dimensional boundary layer

The notion of two-dimensional flow, and *a fortiori* of one-dimensional flow, is an approximation of reality allowing important mathematical simplifications. In addition, it is far easier to conceptualize and conceive a two-dimensional picture than a three-dimensional field. Even with the most modern approaches where the Navier–Stokes equations are solved by a numerical method, the cost of a three-dimensional calculation can be so high that a two-dimensional model must be considered. However, the real world being three-dimensional, this simplification is only justified for a limited number of practical situations (an axisymmetric propulsive nozzle, for example). This is particularly true when the flow separates, the field then nearly always adopting a three-dimensional organization.

The separation phenomenon is linked to the viscous nature of the fluid. In usual aerodynamic flows, where Reynolds numbers are very high (greater then 10^5), viscous effects are normally confined within a thin boundary layer in contact with the obstacle surface. In this boundary layer, the flow velocity changes from the outer value u_e to the value zero imposed by the no-slip condition at the wall. In two-dimensional flow, the evolution of the streamwise velocity component u through the boundary layer is expressed by the relation $u / u_e = f(y)$, where y is the distance from the wall (if the flow is attached, the normal velocity component v plays a minor role). In three dimensions, the velocity vector is no longer constrained to remain in a plane but can rotate of an angle β relative to its direction at the boundary-layer outer edge. It is usual to define the velocity evolution through a three-dimensional boundary layer in a local curvilinear coordinate system, where the longitudinal axis Ox is aligned with the outer velocity vector $\vec{V}_e$, the transverse axis Oz directly perpendicular to Ox and contained in a plane tangent to the obstacle surface, the third axis Oy being normal to the surface (see Figure 1.1). The velocity distribution through the boundary layer is defined by two functions (the normal component v not being considered):

– a profile for the streamwise velocity component: $u / u_e = f(y)$;

– a profile for the transverse (crosswise) component: $w / u_e = g(y)$.

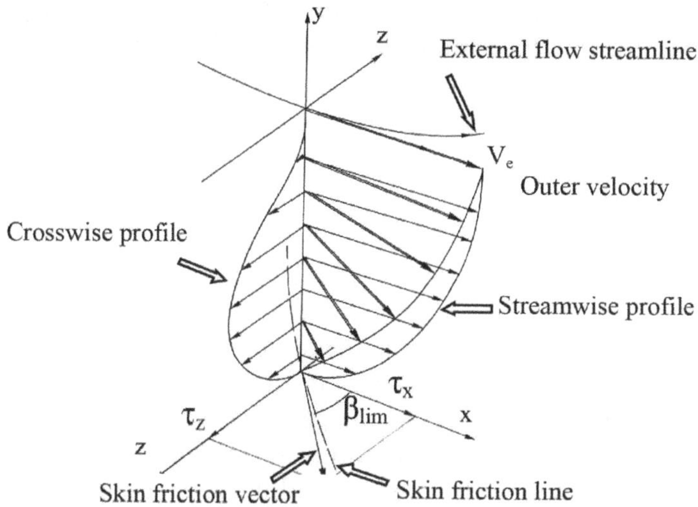

Figure 1.1. *Structure of a three-dimensional boundary layer*

In the chosen coordinate system, w is zero at the boundary-layer outer edge. The w profile characterizes the boundary-layer distortion; that is, the importance of the three-dimensional effect. This effect can be characterized by the distortion angle β defined by the relation:

$$\beta = \tan^{-1}(w/u)$$

On approaching the wall, the components u and w both tend to zero so that we can write the following first-order expansion:

$$u = y\left(\frac{\partial u}{\partial y}\right)_w \quad w = y\left(\frac{\partial w}{\partial y}\right)_w$$

Hence, the limit value for β at the wall is:

$$\beta_{\lim} = \tan^{-1}\left[\left(\frac{\partial w}{\partial y}\right)_w \Big/ \left(\frac{\partial u}{\partial y}\right)_w\right] = \tan^{-1}\ (\tau_z/\tau_x)$$

In the framework of the classical boundary-layer approximation, the components of the skin friction vector at the wall are given by (assuming a Newtonian fluid, which is the case in classical aerodynamics):

$$\tau_x = \mu_w \left(\frac{\partial u}{\partial y}\right)_w \quad \tau_z = \mu_w \left(\frac{\partial w}{\partial y}\right)_w$$

The main property of a three-dimensional flow is the capacity for the boundary layer to develop a transverse component, whereas in two dimension it is constrained to remain in a plane. Under the action of various forces (adverse pressure gradient, in general), a boundary layer having initially two-dimensional properties can develop a transverse profile until a state called *separated* is reached. Existence of a third dimension offers to the flow the possibility to "escape" laterally when it is confronted with an adverse circumstance (most often a pressure rise). Understanding this point has a fundamental importance to thus understand the observed behaviors. In two dimension, the flow must remain in a plane, any possibility to escape in the transverse direction being forbidden for it. Consequently, the gradients associated with separation (compression and deceleration) are much more intense in two dimension than in three dimension. This fact can lead to modeling difficulties (that of turbulence in particular) in great part artificial, since the situations met in reality are rarely two dimensional.

In what follows, we will analyse the separation phenomenon in three-dimensional flows. The two-dimensional case will be reconsidered in Chapter 6 to realize that description of these flows in fact leads to a far greater conceptual complication. We will restrict ourselves to steady flows whose properties (in particular, the velocity field) are independent of time. As in reality flows are most often turbulent – hence unsteady – so we will consider the field resulting from some kind of time averaging for the fluctuating quantities. Although in great part artificial, this assumption is the assumption given by classical measurement techniques, visualization methods (except short-time techniques) and calculations using the time-averaged Navier–Stokes equations (the so-called Reynolds Averaged Navier-Stokes or RANS approach). The considerations that follow can be applied to an unsteady flow by considering an instantaneous picture of the flow field, as the picture given by particle image velocimetry (PIV).

The theory presented later is not a predictive theory since it applies to a vector field given either by experiment or by calculation. It is a descriptive theory used as a support to a consistent description of the field; that is, free of features in contradiction with elementary topological rules.

1.2. Skin friction lines and surface flow pattern

Usually, separation in three-dimensional flows is defined by considering the properties of the flow at the surface of the obstacle. Then, we can introduce concepts allowing a rational definition of separation having the advantage of being experimentally observed. Then, we will see how it is possible to go to the flow field structure in space from its trace on the surface.

Let us consider a body delimited by a surface (S) and define an orthonormal local system $(\vec{n},\vec{i},\vec{j})$ made up of the unit vector $\vec{n}$ normal to the surface and two unit vectors $\vec{i}$ and $\vec{j}$ contained in the plane tangent to (S) at the considered point. Let us designate by (x, y, z) the corresponding coordinates, y being the distance normal to the surface. The fluid flowing past (S) exerts on a surface element dS of (S) a force $\vec{F} = \vec{P}dS$, where $\vec{P}$ is the vector tension that is decomposed in the form:

$$\vec{P} = -p\vec{n} + \vec{\tau}$$

The term $-p\vec{n}$ is the normal action of pressure and $\vec{\tau}$ is the shear stress tangential action. For non-viscous fluids, tension reduces to its normal component. The shear stress $\vec{\tau}$ can be written in the form:

$$\vec{\tau} = \tau_x \vec{i} + \tau_z \vec{j}$$

Thus, in a three-dimensional flow, the wall shear stress is a vector, whereas in two dimensions it is considered as a scalar quantity. The shear stress exerted on the surface (*S*) of the body constitutes a vector field, the problem being to determine the lines of force or trajectories of this vector field. As for streamlines, these lines are defined as the solution curves of the differential system:

$$\frac{dx}{\tau_x(x,z)} = \frac{dz}{\tau_z(x,z)} = dt \qquad [1.1]$$

where t is an integration parameter. We call *skin friction lines* the curves solution of the system [1.1], such curves having the property to be tangent to the local skin friction vector at the contact point. Sometimes, we use the terminology wall streamline (or limit streamline), which is defined as the limit position of a streamline when its distance to the surface tends to zero. In fact, this concept is fictitious since the velocity is null on the surface. As demonstrated above, the limit position of the velocity vector $\vec{V}$ coincides with the shear stress vector, hence:

$$u(x,z,y) = \left(\frac{\partial u}{\partial y}\right)_w y \approx \tau_x y$$

$$w(x,z,y) = \left(\frac{\partial w}{\partial y}\right)_w y \approx \tau_z y$$

Consequently, on approaching the body surface, the streamlines tend to the lines of force of the vector field $\vec{\tau}_w$;

that is, to the skin friction lines. Although the two concepts are equivalent, it is more convenient to use the concept of skin friction lines that can be constructed without difficulty (in principle) from the vectors field $\vec{\tau}_w$ determined from either measurements or calculations. Furthermore, the traces observed on a model covered with a viscous film can be identified with skin friction lines if some precautions are taken in the interpretation of the picture. Figure 1.2 shows the skin friction line pattern visualized with a viscous product on a flattened cylindrical ellipsoid tested in a subsonic wind tunnel.

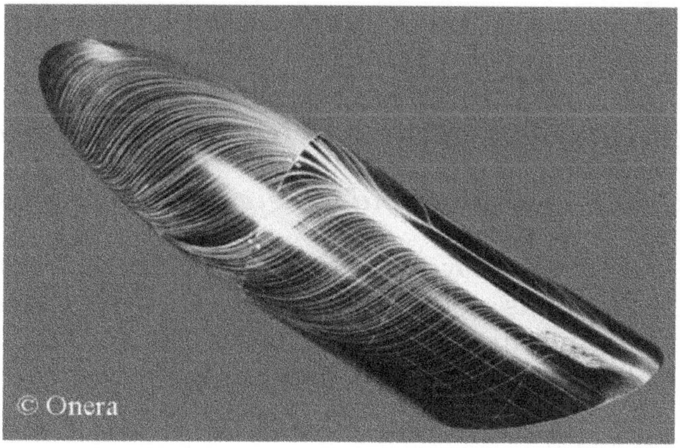

Figure 1.2. *Visualization of the skin friction line pattern over a flattened cylindrical ellipsoid*

We will call surface-flow pattern the shear stress vector field on the surface of a body and the skin friction lines covering the body. The terminology is questionable since there is no flow on the surface but it is commonly used so we will comply with that habit.

Let us consider the vorticity vector $\vec{\Omega}$ whose components are:

$$\omega_x = \frac{\partial v}{\partial z} - \frac{\partial w}{\partial y}, \omega_z = \frac{\partial u}{\partial y} - \frac{\partial v}{\partial x}, \omega_y = \frac{\partial w}{\partial x} - \frac{\partial u}{\partial z}$$

On a wall:

$$\omega_x = -\left(\frac{\partial w}{\partial y}\right)_w, \omega_z = \left(\frac{\partial u}{\partial y}\right)_w, \omega_y = 0$$

Hence the classical results:

The vorticity $\vec{\Omega}_w$ on a body is tangent to the surface and perpendicular to the shear stress vector $\vec{\tau}_w$. The lines of force of vorticity (vortex lines) on the surface are the orthogonal trajectories of the skin friction lines.

1.3. Critical points of the skin friction line pattern

1.3.1. *General solution and the eigenvalue problem*

As seen in the above section, the skin friction lines on a body are solutions of the system [1.1] and, in general, only one skin friction line goes through one point on the surface defined, either by an equation of the form:

$$f(x,z) = Cst.$$

or by the parametric form:

$$x = x(t)$$

$$z = z(t)$$

where t is a parameter such that the skin friction line is traveled by making t vary from $-\infty$ to $+\infty$. In practice, it is exceptional to obtain an analytical definition of the skin friction lines; they must be determined by a step-by-step progress through the skin friction vector field $\vec{\tau}$, as it is done to build streamlines in a velocity field.

The points that do not satisfy the general rule for the tracing of the skin friction lines are the *critical points* (or singular points) of the system [1.1] where simultaneously:

$$\tau_x(x,z)=0,\ \tau_z(x,z)=0$$

To study the skin friction lines behavior near a critical point P_0 of coordinates (x_0, z_0) we make a first-order expansion in the vicinity of P_0:

$$\tau_x=\left(\frac{\partial\tau_x}{\partial x}\right)_{P_0}(x-x_0)+\left(\frac{\partial\tau_x}{\partial z}\right)_{P_0}(z-z_0)$$

$$\tau_z=\left(\frac{\partial\tau_z}{\partial x}\right)_{P_0}(x-x_0)+\left(\frac{\partial\tau_z}{\partial z}\right)_{P_0}(z-z_0)$$

In what follows, it is assumed that the first-order derivatives of the components τ_x and τ_z are non-zero, such a circumstance necessitating higher order expansions. In addition, for the sake of simplicity, the origin of the coordinate system is placed at the critical point P_0, which does not restrict the generality of the problem (this simply consists of making the change of variables $x'=x-x_0$, $z'=z-z_0$; however, to keep simpler notations we will not use primed quantities). At the point P_0, the system [1.1] is written as:

$$\frac{dx}{\left(\frac{\partial\tau_x}{\partial x}\right)_{P_0}x+\left(\frac{\partial\tau_x}{\partial z}\right)_{P_0}z}=\frac{dz}{\left(\frac{\partial\tau_z}{\partial x}\right)_{P_0}x+\left(\frac{\partial\tau_z}{\partial z}\right)_{P_0}z}$$

If λ and μ are two constants, we can write:

$$\frac{dx}{\left(\frac{\partial \tau_x}{\partial x}\right)_{P_0} x+\left(\frac{\partial \tau_x}{\partial z}\right)_{P_0} z}=\frac{dz}{\left(\frac{\partial \tau_z}{\partial x}\right)_{P_0} x+\left(\frac{\partial \tau_z}{\partial z}\right)_{P_0} z}$$

$$=\frac{\lambda dx+\mu dz}{\lambda\left[\left(\frac{\partial \tau_x}{\partial x}\right)_{P_0} x+\left(\frac{\partial \tau_x}{\partial z}\right)_{P_0} z\right]+\mu\left[\left(\frac{\partial \tau_z}{\partial x}\right)_{P_0} x+\left(\frac{\partial \tau_z}{\partial z}\right)_{P_0} z\right]}$$

A solution of the system is looked for by writing the above expression in the form of a logarithmic derivative:

$$\frac{\lambda dx+\mu dz}{\lambda\left[\left(\frac{\partial \tau_x}{\partial x}\right)_{P_0} x+\left(\frac{\partial \tau_x}{\partial z}\right)_{P_0} z\right]+\mu\left[\left(\frac{\partial \tau_z}{\partial x}\right)_{P_0} x+\left(\frac{\partial \tau_z}{\partial z}\right)_{P_0} z\right]}=\frac{d(\lambda dx+\mu dz)}{S(\lambda x+\mu z)}$$

where S is a constant. If this form is possible, then by putting:

$$\frac{d(\lambda dx+\mu dz)}{S(\lambda x+\mu z)}=-dt$$

we have:

$$\lambda x+\mu z=A\exp(-St)$$

where t is the integration variable and A is a constant.

For the logarithmic form to be possible, we must simultaneously have:

$$\left[\left(\frac{\partial \tau_x}{\partial x}\right)_{P_0}-S\right]\lambda+\left(\frac{\partial \tau_z}{\partial x}\right)_{P_0}\mu=0$$

$$\left(\frac{\partial \tau_x}{\partial z}\right)_{P_0}\lambda+\left[\left(\frac{\partial \tau_z}{\partial z}\right)_{P_0}-S\right]\mu=0$$

This linear algebraic system, homogeneous for S, admits non-trivial solutions if its determinant is non-zero; hence the condition on S:

$$\begin{vmatrix} \left(\dfrac{\partial \tau_x}{\partial x}\right)_{P_0} - S & \left(\dfrac{\partial \tau_z}{\partial x}\right)_{P_0} \\ \left(\dfrac{\partial \tau_x}{\partial z}\right)_{P_0} & \left(\dfrac{\partial \tau_z}{\partial z}\right)_{P_0} - S \end{vmatrix} = 0$$

The skin friction lines behavior in the vicinity of the critical point is dictated by the nature of the solutions for the above second-degree equations for S. Before continuing the discussion, let us introduce the Jacobian matrix:

$$F = \begin{vmatrix} \left(\dfrac{\partial \tau_x}{\partial x}\right)_{P_0} & \left(\dfrac{\partial \tau_z}{\partial x}\right)_{P_0} \\ \left(\dfrac{\partial \tau_x}{\partial z}\right)_{P_0} & \left(\dfrac{\partial \tau_z}{\partial z}\right)_{P_0} \end{vmatrix}$$

and put:

$$p \equiv -\text{ trace of } F = -\left[\left(\frac{\partial \tau_x}{\partial x}\right)_{P_0} + \left(\frac{\partial \tau_z}{\partial z}\right)_{P_0}\right]$$

$$q \equiv \text{determinant of } F = \left(\frac{\partial \tau_x}{\partial x}\right)_{P_0}\left(\frac{\partial \tau_z}{\partial z}\right)_{P_0} - \left(\frac{\partial \tau_z}{\partial x}\right)_{P_0}\left(\frac{\partial \tau_x}{\partial z}\right)_{P_0}$$

The equation giving the eigenvalues S_1 and S_2 can be written in the condensed form as:

$$S^2 + pS + q = 0 \qquad [1.2]$$

Hence the solutions:

$$S_{1,2} = \frac{-p \pm \sqrt{p^2 - 4q}}{2}$$

To any eigenvalue $S_{1,2}$ corresponds an *eigenvector* of components $\left[\lambda_{1,2}, \mu_{1,2}\right]$. The eigenvalues S_1 and S_2 depend only on p and q, which means that the Jacobian matrix is the only function of the local properties of the surface flow at the critical point, such properties (the derivatives of the shear stress components at P_0) resulting from a Navier–Stokes calculation or experiments. The above classical analysis allows us to solve the indeterminate form taken by the system [1.1] at the critical point when the shear stress vector vanishes. It allows the construction of the skin friction lines in the vicinity of P_0. The study of the solution near a critical point leads to an eigenvalue problem (here S_1 and S_2) similar to the problem encountered in linear stability analyses. The roots S_1 and S_2 are the eigenvalues of the Jacobian matrix and the associated solutions the eigenvectors of this matrix.

To S_1 and S_2 correspond an infinity of solutions for which λ and μ are proportional, we can choose:

$$\lambda_1 = -\left(\frac{\partial \tau_z}{\partial x}\right)_{P_0}, \qquad \mu_1 = \left(\frac{\partial \tau_x}{\partial x}\right)_{P_0} - S_1 \qquad \text{for } S_1$$

and

$$\lambda_2 = \left(\frac{\partial \tau_z}{\partial z}\right)_{P_0} - S_2, \qquad \mu_2 = -\left(\frac{\partial \tau_x}{\partial z}\right)_{P_0} \qquad \text{for } S_2$$

Thus, the solution determining the skin friction lines will be of the form:

$$\lambda_1 x + \mu_1 z = A_1 \exp(-S_1 t)$$

$$\lambda_2 x + \mu_2 z = A_2 \exp(-S_2 t) \qquad [1.3]$$

where A_1 and A_2 are up to now undetermined constants. The solution of system [1.3] in term of (x,z) gives:

$$x(t) = \frac{A_1\mu_2 \exp(-S_1 t) - A_2\mu_1 \exp(-S_2 t)}{\lambda_1\mu_2 - \lambda_2\mu_1}$$

$$z(t) = -\frac{A_1\lambda_2 \exp(-S_1 t) - A_2\lambda_1 \exp(-S_2 t)}{\lambda_1\mu_2 - \lambda_2\mu_1} \qquad [1.4]$$

The slope of the skin friction lines at point P_0 is given by:

$$\frac{dz}{dx} = -\frac{\lambda_2 - \dfrac{S_2 A_2}{S_1 A_1}\lambda_1 \exp[-(S_2 - S_1)t]}{\mu_2 - \dfrac{S_2 A_2}{S_1 A_1}\mu_1 \exp[-(S_2 - S_1)t]}$$

The shape of the solutions depends on the nature (real or complex) and sign of the eigenvalues at point P_0. The solution behavior can be discussed by considering the plane $[p,q]$ where the parabola (P) of equation (see Figure 1.3) is plotted:

$$q = \frac{p^2}{4}$$

The surface flow at P_0 has an "image" in this plane so that according to the location of this image relative to the parabola (P) and to the axes $p = 0$ and $q = 0$, the skin friction lines will have specific behaviors.

Considering the expression of vorticity given earlier, we immediately conclude:

Vorticity being zero where the shear stress vector vanishes, the critical points of the skin friction line pattern are also critical points for the vorticity-line pattern on the wall.

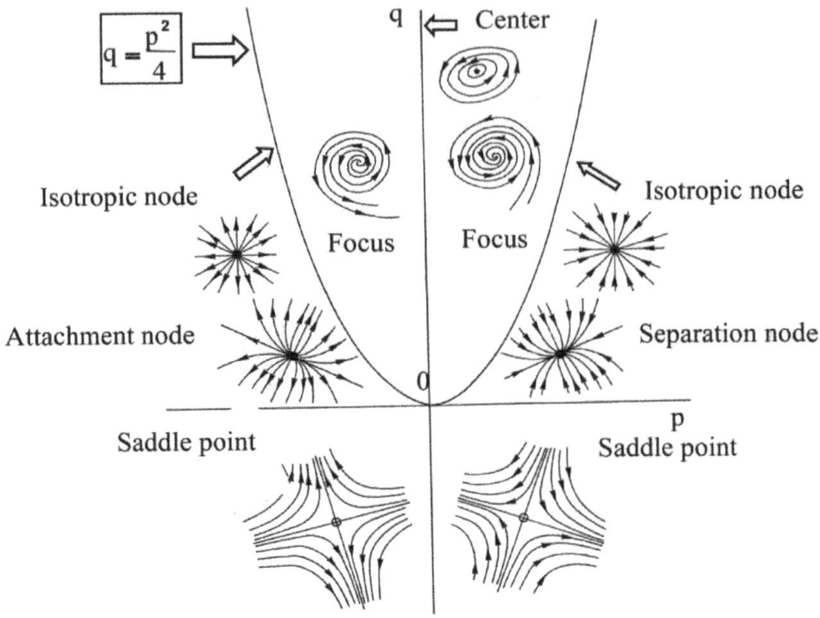

Figure 1.3. *Critical points classification in the plane* $[p,q]$

1.3.2. *The different critical points*

1.3.2.1. *Critical point of the node type*

In the present case, the two roots S_1 and S_2 are real, distinct with the same sign, this occurs if:

$$p^2 - 4q > 0,\ q > 0$$

The representative point in the plane $[p,q]$ is below the parabola (the roots being real) and above the axis $q = 0$ (since they have the same sign).

Let us examine the solution behavior in the vicinity of P_0 by assuming $S_2 > S_1$. The relations [1.4] show that all the skin friction lines go through P_0 which is reached when $t \to \infty$. Indeed, we can easily verify that $x(t)$ and $z(t)$ simultaneously tend to 0 when $t \to \infty$. In addition, all the skin friction lines have the same slope at P_0, which is given by:

$$\left.\frac{dz}{dx}\right|_{P_0} = -\frac{\lambda_2}{\mu_2}$$

Let us consider the lines corresponding to $A_1 = 0$ and $A_2 = 0$. If $A_1 = 0$:

$$x(t) = -\frac{A_2 \mu_1 \exp(-S_2 t)}{\lambda_1 \mu_2 - \lambda_2 \mu_1} \qquad z(t) = \frac{A_1 \lambda_2 \exp(-S_2 t)}{\lambda_1 \mu_2 - \lambda_2 \mu_1}$$

Then: $\left.\dfrac{dz}{dx}\right|_{P_0} = -\dfrac{\lambda_1}{\mu_1}$

If $A_2 = 0$:

$$x(t) = \frac{A_1 \mu_2 \exp(-S_1 t)}{\lambda_1 \mu_2 - \lambda_2 \mu_1} \qquad z(t) = -\frac{A_1 \mu_2 \exp(-S_1 t)}{\lambda_1 \mu_2 - \lambda_2 \mu_1}$$

and:

$$\left.\frac{dz}{dx}\right|_{P_0} = -\frac{\lambda_2}{\mu_2}$$

Consequently, at P_0 all the skin friction lines are tangent to the line corresponding to $A_2 = 0$, except the line for which $A_1 = 0$. In other words, at the critical point all the skin friction lines have a common tangent, except for one of them. The surface flow in the vicinity of P_0 is represented in Figure 1.4(a). The two eigenvectors are orthogonal for the node shown in Figure 1.4(b).

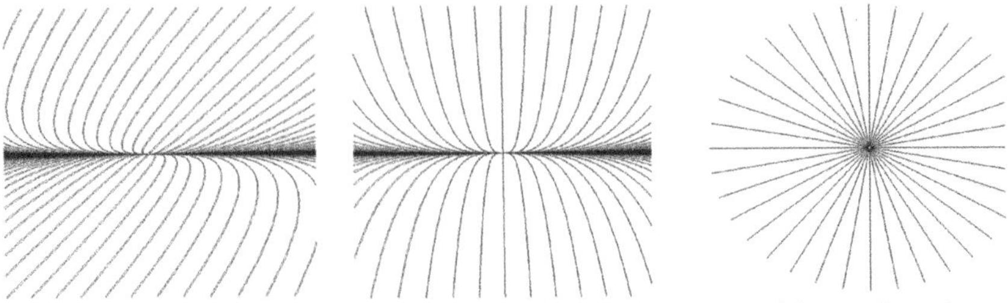

Figure 1.4. *Critical point of the node type* ($S_1 = 4, S_2 = 0$)

Such a critical point is called a *node*. When t varies from $-\infty$ to $+\infty$, the skin friction lines run starting from the critical point (if A_1 and A_2 are negative). Such a node is an *attachment node.* If the skin friction lines run toward the critical point (A_1 and A_2 are positive), the node is a *separation node*. If A_1 and A_2 are negative, then p is positive so that the representative points in the plane $[p,q]$ are on the left-hand side of the axis $q = 0$. For A_1 and A_2 positive, the image points are on the right-hand side of this axis.

Figure 1.5 shows a surface flow visualization of an attachment node at the nose of a blunted body having the shape of a flattened ellipsoid.

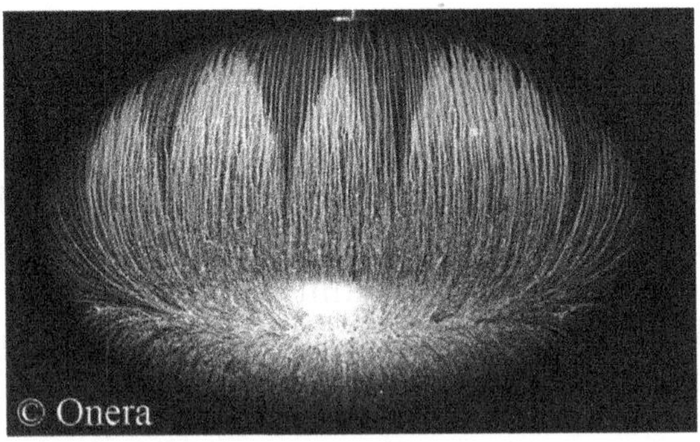

Figure 1.5. *Attachment node at the nose of a blunt body*

Special case: $S_1 = S_2$

The skin friction lines are defined by:

$$x(t) = \frac{A_1\mu_2 - A_2\mu_1}{\lambda_1\mu_2 - \lambda_2\mu_1}\exp(-S_1 t)$$

$$z(t) = -\frac{A_1\lambda_2 - A_2\lambda_1}{\lambda_1\mu_2 - \lambda_2\mu_1}\exp(-S_1 t)$$

Hence:

$$\frac{dz}{dx} = \frac{A_1\lambda_2 - A_2\lambda_1}{A_2\mu_1 - A_1\mu_2}$$

All the skin friction lines go through the critical point P_0 where they all have a different slope. Such a critical point is called an *isotropic node*, the corresponding pattern being shown in Figure 1.4(c). According to the sign of S_1, the skin friction lines are traveled either from P_0 (case of an attachment isotropic node) or to P_0 (case of a separation isotropic node). In the plane $[p,q]$, isotropic nodes (which correspond to double solutions of the eigenvalue equation) have images located on the parabola (P).

1.3.2.2. *Critical point of the saddle point type*

Now, the two roots S_1 and S_2 are real, with opposite signs and such that (to fix ideas): $S_1 < 0 < S_2$. Let us recall the following expressions for the solution curves:

$$x(t) = \frac{A_1\mu_2}{\lambda_1\mu_2 - \lambda_2\mu_1}\exp(-S_1 t) - \frac{A_2\mu_1}{\lambda_1\mu_2 - \lambda_2\mu_1}\exp(-S_2 t)$$

$$z(t) = \frac{-A_1\lambda_2}{\lambda_1\mu_2 - \lambda_2\mu_1}\exp(-S_1 t) + \frac{A_2\lambda_1}{\lambda_1\mu_2 - \lambda_2\mu_1}\exp(-S_2 t)$$

which can be written in the form:

$$x(t) = \frac{1}{\lambda_1\mu_2 - \lambda_2\mu_1} \exp(-S_1 t)\left\{A_1\mu_2 - A_2\mu_1 \exp\left[-(S_2 - S_1)t\right]\right\}$$

$$z(t) = \frac{-1}{\lambda_1\mu_2 - \lambda_2\mu_1} \exp(-S_1 t)\left\{A_1\lambda_2 - A_2\lambda_1 \exp\left[-(S_2 - S_1)t\right]\right\}$$

The following behaviors can be verified:

when: $t \to +\infty, \exp(-S_1 t) \to +\infty \quad \text{and} \quad \exp(-S_2 t) \to 0$

Hence $x(t)$ and $z(t)$ tend to $+\infty$.

when: $t \to -\infty, \exp(-S_1 t) \to 0 \quad \text{and} \quad \exp(-S_2 t) \to +\infty$

So that $x(t)$ and $z(t)$ still tend to $+\infty$.

In the general case, *the critical point is never reached.* If the parameter t varies from $-\infty$ to $+\infty$ a skin friction line runs from a point infinitely far from P_0 when tending to P_0; thereafter, it continues to a point infinitely far when moving away from P_0.

Let us consider the case: $A_1 = 0$ or $A_2 = 0$.

If $A_1 = 0$:

$$x(t) = -\frac{A_2\mu_1}{\lambda_1\mu_2 - \lambda_2\mu_1} \exp(-S_2 t)$$

$$z(t) = \frac{A_2\lambda_1}{\lambda_1\mu_2 - \lambda_2\mu_1} \exp(-S_2 t)$$

Thus, simultaneously $x(t)$ and $z(t) \to 0$ when $t \to +\infty$. The solution curve passes through P_0 where its slope is:

$$\frac{dz}{dx} = -\frac{\lambda_1}{\mu_1}$$

If $A_2 = 0$:

$$x(t) = \frac{A_1\mu_2}{\lambda_1\mu_2 - \lambda_2\mu_1}\exp(-S_1 t)$$

$$z(t) = \frac{-A_1\mu_2}{\lambda_1\mu_2 - \lambda_2\mu_1}\exp(-S_1 t)$$

We see that $x(t)$ and $z(t) \to 0$ when $t \to +\infty$: the critical point is reached, the slope at P_0 being:

$$\frac{dz}{dx} = -\frac{\lambda_2}{\mu_2}$$

Only two skin friction lines run through P_0; all the other skin friction lines avoid the critical point adopting the shape of a hyperbolic curve, as shown in Figure 1.6(a).

a) eigenvectors $(\lambda_1 = -6, \lambda_2 = 1)$ b) orthogonal eigenvectors $(\lambda_1 = 0, \lambda_2 = 100)$

Figure 1.6. *Critical point of the saddle point type* $(S_1 = -1,\ S_2 = 1)$

Such a critical point is called a *saddle point*. For the saddle point drawn in Figure 1.6(b), the two eigenvectors are orthogonal. If the coefficient q in equation [1.2] is negative, the roots S_1 and S_2 have different signs. The representative point in the plane $[p,q]$ lies below the axis $q=0$ in Figure 1.3. Figure 1.7 shows a saddle point observed in the surface pattern of a flow separating in front of an obstacle.

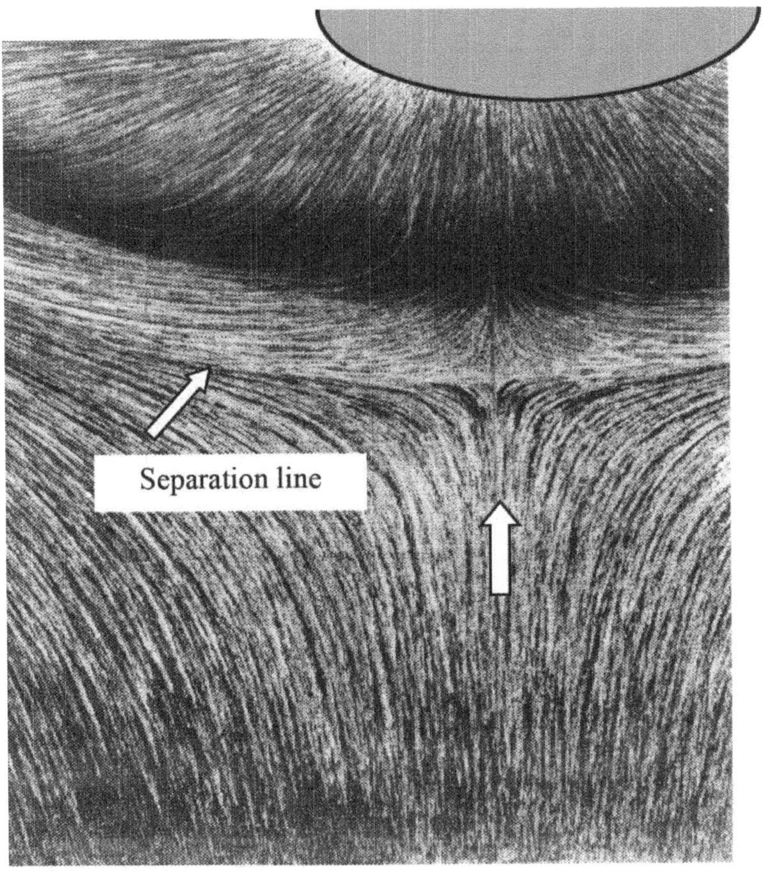

Figure 1.7. *Saddle point linked to separation in front of a cylindrical obstacle ([EAS 68])*

1.3.2.3. *Critical point of the focus type*

If the discriminant in equation [1.2] is negative, the eigenvalues are a complex conjugate. Then we can write:

$$S_1 = S' + iS''$$

$$S_2 = S' - iS''$$

S_1 correspond to:

$$\lambda_1 = \lambda_1' + i\lambda_1'' \quad \text{and} \quad \mu_1 = \mu_1' + i\mu_1''$$

and S_2 to:

$$\lambda_2 = \lambda_2' + i\lambda_2'' \quad \text{and} \quad \mu_2 = \mu_2' + i\mu_2''$$

Let us consider the root $S = S' + iS''$ to which are associated $\lambda = \lambda'' + i\lambda''$ and $\mu = \mu' + i\mu''$, the index 1 (or 2) being omitted for sake of simplicity. The integration constant A being written in the form:

$$A = \exp(\alpha + i\beta)$$

We have from the system [1.3]:

$$(\lambda' + i\lambda'')x + (\mu' + i\mu'')z = \exp\left[-(S' + iS'')t + \alpha + i\beta\right]$$

Hence, by separating the real and imaginary parts:

$$\lambda' x + \mu' z = \cos(S'' t - \beta)\exp(-S' t + \alpha)$$

$$\lambda'' x + \mu'' z = -\sin(S'' t - \beta)\exp(-S' t + \alpha)$$

Then:

$$x(t) = \frac{\mu'' \cos(S'' t - \beta)\exp(-S' t + \alpha) + \mu' \sin(S'' t - \beta)\exp(-S' t + \alpha)}{\lambda' \mu'' - \lambda'' \mu'}$$

$$z(t)=\frac{\lambda''\cos(S''t-\beta)\exp(-S't+\alpha)+\lambda'\sin(S''t-\beta)\exp(-S't+\alpha)}{\lambda'\mu''-\lambda''\mu'}$$

If we assume that S' is strictly positive $\left(S'>0\right)$, then it is easy to verify that $x(t)$ and $z(t)$ tend to zero when $t\to+\infty$. The critical point P_0 is reached by all the solution curves. Let us examine the slope:

$$\frac{dz}{dx}=-\frac{(S''\mu'-S'\mu'')\cos(S''t-\beta)-(S''\mu''+S'\mu')\sin(S''t-\beta)}{(S''\lambda'-S'\lambda'')\cos(S''t-\beta)-(S''\lambda''+S'\lambda')\sin(S''t-\beta)}$$

At P_0, dz/dx takes different values according to the value of the constant β: the skin friction lines have no common tangent. Because of the presence of periodic functions damped by exponentials, the skin friction lines end at the critical point P_0 after spiralling around it, as shown in Figure 1.8(a). Such a critical point is called a *focus*.

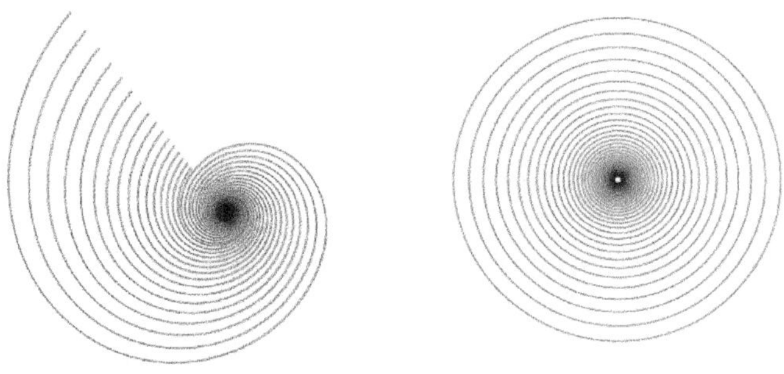

a) critical point for $S_{1,2}=1\pm 4i$ b) degenerated focus or center $\left(S_{1,2}=\pm 4i\right)$

Figure 1.8. *Critical point of the focus type*

A focus around which the solution curves roll up is sometimes called a stable focus. In fluid mechanics, such behavior is associated with a vortex, or more exactly with a vortical surface resulting from separation.

If S' is strictly negative $(S' < 0)$, $x(t)$ and $z(t) \to 0$ when $t \to -\infty$ and $x(t)$ and $z(t) \to +\infty$ when $t \to -\infty$: the solution curves unroll from the critical point that is then called an instable focus. The denominations stable or instable focus pertain to the dynamic systems theory. We should avoid deducing any conclusion relative to the hydrodynamic stability of a flow (that is a vortex breakdown). In the case of a focus $q > p^2 / 4$, so that the representative point in the $[p,q]$ plane is inside the parabola of Figure 1.3.

In the case $S' = 0$, we have:

$$x(t) = \frac{\exp(\alpha)\left[\mu''\cos(S''t - \beta) + \mu'\sin(S''t - \beta)\right]}{\lambda'\mu'' - \lambda''\mu'}$$

$$x(t) = \frac{\exp(\alpha)\left[\lambda''\cos(S''t - \beta) + \lambda'\sin(S''t - \beta)\right]}{\lambda'\mu'' - \lambda''\mu'}$$

The skin friction lines are closed curves having an elliptical shape that encircles a limit curve reducing to a point coincident with $.\alpha \to -\infty.$ For the other solution lines, P_0 is never reached. Such a point, which is a special kind of degenerated focus, is called a *center* (see Figure 1.8(b)). In the plane $[p,q]$, centers correspond to points located on the axis $p = 0$. Figure 1.9 shows two foci forming on the rear window of an automobile.

Other critical points corresponding to special situations in the plane $[p,q]$ are considered in Chapter 6 devoted to the two-dimensional case reconsidered within the framework of the critical point theory.

The various critical points used to study the topology of separated three-dimensional flows are represented in Figure 1.10. The sense along which the skin friction lines are run allows giving the critical point physical signification by linking it either to an attachment or a separation. These notions are clarified in the following chapter.

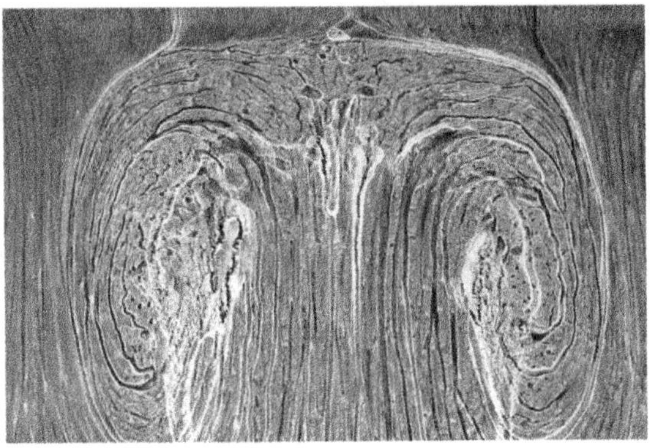

Figure 1.9. *Two-foci combination on an automobile rear window*

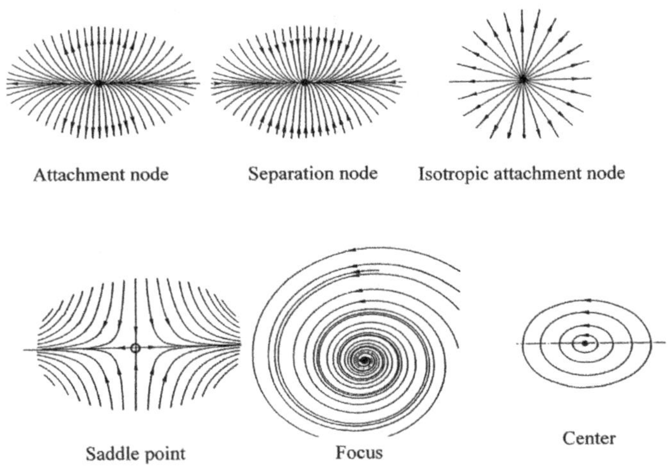

Figure 1.10. *The main critical points used for the topological description of three-dimensional flows*

1.4. Critical points of the wall vorticity lines

The critical points of the shear stress and vorticity fields are coincident since they simultaneously vanish as seen in section 1.2. Hence, the previous analysis can be applied to the vector field $\vec{\Omega}_w$, the orthogonal property of the skin friction and vorticity lines allowing us to deduce the nature of the critical point for the vector field $\vec{\Omega}_w$ from that of the vector field $\vec{\tau}_w$. Consequently, the following results are given without demonstration. As shown in Figure 1.11:

– to a node for $\vec{\tau}_w$ corresponds a center for $\vec{\Omega}_w$. If the node is isotropic, the curves constituting the center for $\vec{\Omega}_w$ are circles;

– to a saddle point for $\vec{\tau}_w$ corresponds for $\vec{\Omega}_w$ a saddle point having undergone a rotation;

– to a focus for $\vec{\tau}_w$ corresponds a focus for $\vec{\Omega}_w$;

– to a center for $\vec{\tau}_w$ corresponds a node for $\vec{\Omega}_w$.

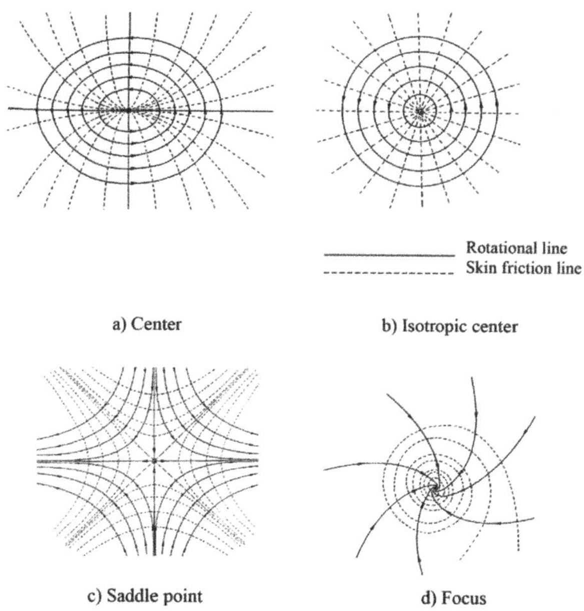

Figure 1.11. *Critical points for the rotational field at the wall*

2

Separation Streamsurfaces and Vortex Structures

The critical points are generalized to the flow field, which allows introducing the basic concepts of separation (or attachment) lines and separation (or attachment) surfaces. It is shown that the wrapping of separation surfaces leads to the formation of vortical surfaces (or vortices) taking the shape of horseshoe vortices or tornado-like vortices.

2.1. Generalization to the flow field and three-dimensional critical points

The preceding considerations, which apply to any vector field, can be extended to a three-dimensional flow. For a steady regime (not time dependent), the streamlines are defined by the differential equations:

$$\frac{dx}{u(x,y,z)} = \frac{dy}{v(x,y,z)} = \frac{dz}{w(x,y,z)}$$

and we examine their behavior in the vicinity of a critical point where:

$$u = v = w = 0$$

by means of a first-order expansion of the velocity components. The equations thus obtained can be integrated by a linear combination with constant coefficients λ, μ and ν, which are determined in such a way that the equation takes the form of a logarithmic derivative:

$$\frac{d(\lambda x+\mu y+\nu z)}{S(\lambda x+\mu y+\nu z)}=dt$$

The critical points are of a different nature according to the eigenvalues that are solutions of a third-order algebraic equation. We will not develop this discussion here, which would lead to a rather complex mathematical development which is not useful for what follows.

1) If the three roots, S_1, S_2 and S_3, are real and of the same sign, the critical point P_0 is a node: all the solution curves that enter a little sphere surrounding P_0 converge to this point.

2) If S_1, S_2 and S_3 are real with different signs, the critical point P_0 is a saddle point.

3) If S_1 is real and S_2 and S_3 are complex conjugates, the critical point is a focus.

Figure 2.1 shows a three-dimensional focus forming what is called a *tornado-like vortex*. The trace of the structure on the body surface is a focus around which the skin friction lines roll up. The outer field streamlines, which are attracted by the structure, spiral around a special line – or axis – going through the focus F. These lines contract near the structure's axis as they lift off from the surface, resulting in a winding and ascending motion. Figure 2.2 is a water tunnel visualization of the vortex generated by the suction effect of an air intake when the plane is on the tarmac with the engine on. The vortex swallowed by the air intake is fed by the ground boundary layer; hence, a risk of debris ingestion by the engine exists.

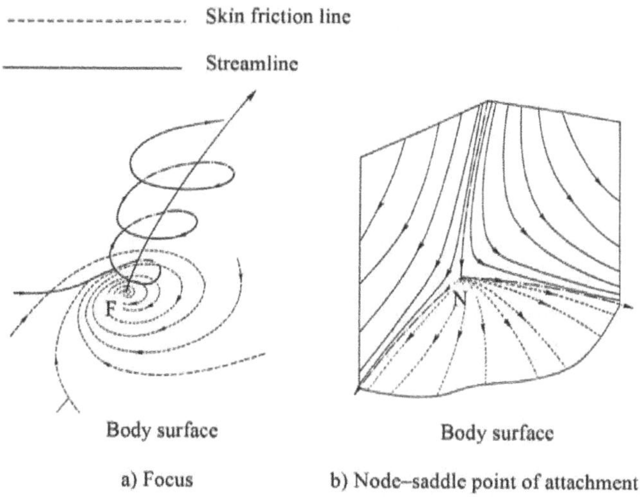

Figure 2.1. *Three-dimensional focus and node*

Figure 2.2. *Water tunnel visualization of the vortex generated by the suction effect of an air intake near the ground. For a color version of this figure see www.iste.co.uk/delery/3Dflow.zip*

Figure 2.3 shows a three-dimensional node that is the combination of a saddle point in the flow and a node on the body surface. Such a structure corresponds to attachment – or stagnation – at the nose of a blunt body, as shown in Figure 2.3(a). The skin friction lines emanating from the node N wrap the whole body. When approaching the body, the outer field streamlines close to the nose rapidly turn to circle the obstacle. There is only one stagnation point on the body so that all the streamlines (except one) avoid it, the outer field containing a half-saddle point. By changing the flow direction along the streamlines, we obtain a separation point at the body downstream extremity (see Figure 2.3(b)).

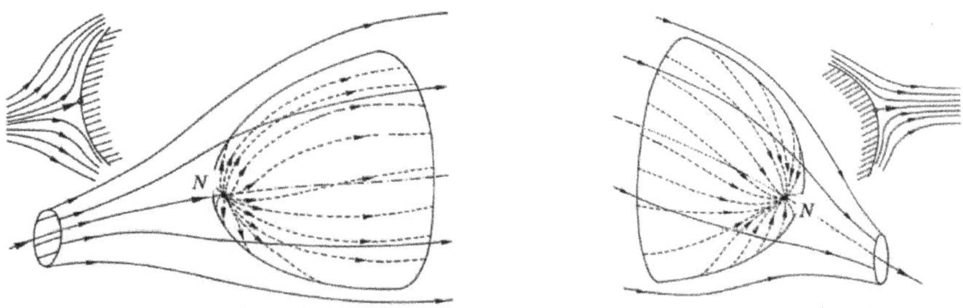

a) Attachment (stagnation) node b) Separation node

Figure 2.3. *Three-dimensional node*

It is possible to conceive a structure made of a bubble of fluid that would occupy the place of the body. As shown in Figure 2.4(a), the flow field then contains a streamsurface (Σ) with a node N from which the streamlines constituting (Σ) emanate. Behind N, the surface (Σ) is approached by a stream flowing back when reaching N to constitute a recirculating flow, two half three-dimensional saddle points existing before and behind N. The same organization can be imagined at the closure of the separation bubble as shown in Figure 2.4(b).

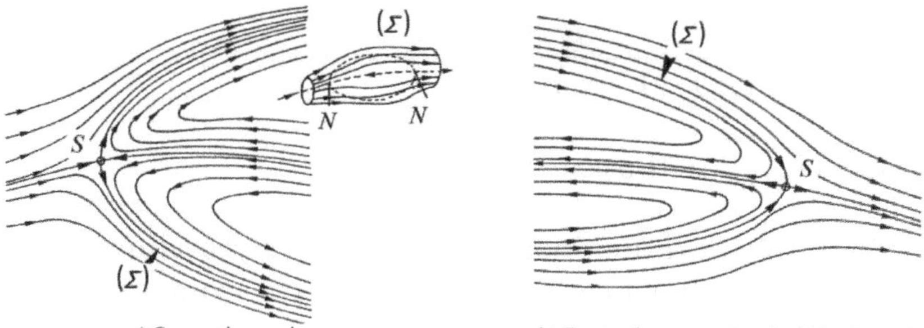

a) Separation region b) Reattachment region (bubble closure)

Figure 2.4. *Recirculation bubble in a flow field*

Existence of a stagnation point inside a flow field is rarely observed, the stagnation points being most often located on a material body. From a certain point of view, a material surface is a distribution of points where velocity is zero because of the no-slip condition, hence of singularities or critical points. As seen in section 1.2, when the distance to a wall tends to zero, streamlines continuously tend to skin friction lines so that intuitively reasoning, we can conceive that the critical points of the outer field coincide with those of the surface. Nodes and saddle points of the flow field coincide with the singularities of the skin-friction line pattern. We will see that a node in the skin-friction line pattern is associated with a saddle point in the contiguous flow field and that a saddle point on the surface is associated with a node in the outer flow.

The critical point theory also applies to the vector field obtained by performing a projection of the three-dimensional field in a plane. Perception of three-dimensional objects being difficult, projection of the vector field in well-chosen planes is commonly used in flow analysis. Let us consider the projection of the vector $\vec{V}$ in the plane (P) and designate by V_1 and V_2 the components of the projection $\vec{V}_P$ of $\vec{V}$ in (P). The field made of vectors $\vec{V}_P$ constitutes a regular vector field whose lines of force (trajectories) are solutions of the system:

$$\frac{dx}{V_1(x,y,z)} = \frac{dy}{V_2(x,y,z)}$$

where (x, y) are the coordinates in (P).

Such lines are distinct from the true streamlines of the three-dimensional flow, except if (P) is a symmetry plane (then the component normal to (P) is zero). The above system being identical to the system defining the skin friction lines, the same discussion can be had by considering the points where $V_1 = V_2 = 0$ and introducing critical points such as nodes, foci and saddle points in the pattern of the pseudo-streamlines. Examination of the projected velocity field (or of its associated pseudo-streamlines) can lead to observations that should be considered with circumspection as far as their physical significance is concerned: the field thus obtained depends on the projection plane; it does not give an *objective* picture of the aerodynamic field.

Except for special situations (a symmetry plane, for example), the denomination "streamline", frequently used for the projected field, is misleading since the velocity $\vec{V}$ has a component normal to the projection plane. The true streamlines, which reflect the real flow dynamics, are different from the lines observed in the projection plane. The term "trajectory", which has a precise and specific meaning, should be avoided, as well as the denomination "lines of force" (of the projected field) that belongs to electromagnetism. Although not satisfactory, the denomination *pseudo-streamlines* will be used in the following.

2.2. Separation and attachment lines

The saddle point plays a key role in the analysis of separated flows. Let us recall that in the vicinity of such a point, all the skin friction lines avoid the point except two that go through the critical point. Such lines will be called *separation skin friction lines* or more briefly *separation lines*, although they can be associated with an attachment process. Their mathematical denomination as *separatrices* will be avoided because they are not used in fluid mechanics. This will lead to some ambiguity because a separation line can be linked to the separation of the flow and its attachment. This difficulty could be avoided by replacing flow separation by flow detachment (in French *décollement*), with the drawback of changing a widely spread habit.

Let us consider the situation shown in Figure 2.5(a). The skin friction lines coming from upstream and located on each side of the separation line (S_2) divide themselves into two families when approaching the saddle point S: one family turns "to the right" whereas the other family turns "to the left". After being separated, the skin friction lines of the two families follow the second separation line (S_1) converging toward it without any contact between these lines and (S_1). The converging motion toward (S_1) can be more or less rapid according to the circumstances imposed by the outer flow, the pressure gradient in particular. Frequently, the skin friction lines run along the separation line over a great distance without any converging motion. The behavior shown in Figure 2.5(a) is associated with the separation (or detachment) phenomenon, as it will be seen below. The visualization in Figure 2.7 shows such a behavior in front of a cylindrical obstacle.

A similar behavior is observed on the other side of the separation line (S_1), where the skin friction lines flow toward the saddle point and turn either to the right or to the left according to their location with respect to (S_2). Thereafter, the two families thus separated converge toward the separation line (S_1). In another opposite circumstance (shown in Figure 2.5(b)), the skin friction lines first run along the separation line (S_1) and then move away from it when approaching the saddle point S, the two families existing on each side of (S_1). Here too, the divergence motion from (S_1) can be more or less rapid, according to external conditions. Such a pattern is associated with flow attachment.

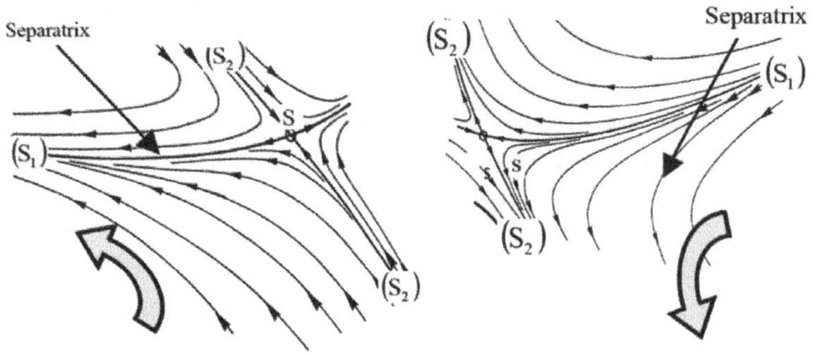

a) Separation (separatrix) line of detachment

b) Separation line of attachment

Figure 2.5. *Skin friction line of separation (separatrix)*

The flow behavior above the wall can be interpreted by a simplified and rather crude reasoning. Let us consider in Figure 2.6(a) a streamtube, with a rectangular cross-section, limited by the two skin friction lines (f_1) and (f_2) and the two streamlines (l_1) and (l_2). Let us designate by n the width of the tube, by h its height, and by $\bar{\rho}$ and $\bar{V}$ the mean values of the density and velocity, respectively, over the streamtube section. The mass flow through the streamtube is:

$$q_m = \bar{\rho} n h \bar{V}$$

Let us consider the case when the skin friction lines converge when approaching the saddle point S so that the distance n between (f_1) and (f_2) becomes smaller and smaller. Consequently, since the mean velocity and density remains finite and as the streamtube mass flow q_m is constant (the flow is assumed steady), its height increases:

$$h = \frac{q_m}{\overline{\rho} n \overline{V}}$$

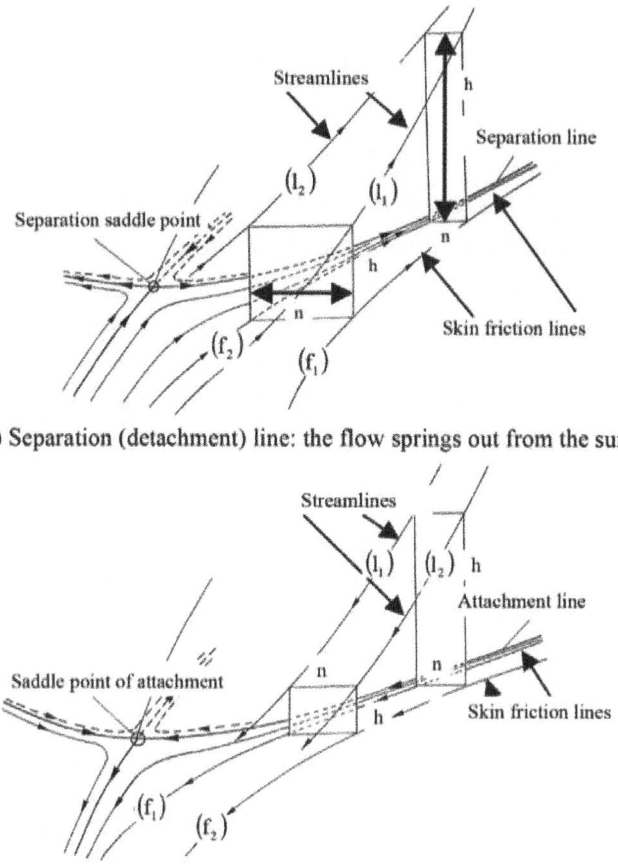

a) Separation (detachment) line: the flow springs out from the surface

b) Separation line of attachment: the flow dives toward the surface

Figure 2.6. *Flow behavior in the vicinity of a separation line*

Thus, when approaching the separation line (S_1), the streamlines move away from the wall: they "take off". The layer of fluid in contact with the wall expands in the vertical direction, this dilatation being more ample as the converging motion toward (S_1) is faster. This behavior is typical of what is called separation (or detachment). In such a phenomenon, the regions of fluid in contact with the wall, where vorticity is concentrated (the boundary layers), "lift off" from the wall and spring up into the flow region considered as non-viscous. This motion, frequently spectacular, is at the origin of the formation of vortices through which vorticity invades the part of the flow said to be "sound".

In the opposite situation, shown in Figure 2.6(b), the skin friction lines (f_1) and (f_2) move away from (S_1) so that the streamtube width n increases. The mean quantities $\bar{\rho}$ and $\bar{V}$ remaining finite, the height h decreases so that the streamtube "crashes" on the wall in the vicinity of C and the outer fluid dives to the body surface. Such a behavior is termed *attachment*. In this case, the outer "sound" flow feeds the dissipative layer, vorticity being produced by the gradients resulting from the joint action of viscosity and intense velocity retardation. Sometimes, the designation reattachment is used, which implies the existence of a previous separation, the flow having separated being again in contact with the wall. Such a special circumstance met in two-dimensional flow is exceptional in three-dimensional flows.

2.3. Streamsurfaces of separation and attachment

A skin friction line of separation is the trace on the body surface of a particular streamsurface called *separation streamsurface*. Such a surface (Σ_1), supported by (S_1), is represented in Figures 2.7(a) and (b) in the case of an effective separation (detachment). The streamlines constituting the separation surface have their origin at the node N belonging to the outer flow (that is not the surface flow), which coincides with the saddle point S of the surface flow pattern. As seen in section 2.1, the set [N + S] constitutes a three-dimensional critical point where the velocity is zero. All the streamlines of the streamsurface (Σ_1) flow from N. Close to the wall, these lines "run" along (S_1) without any contact with it.

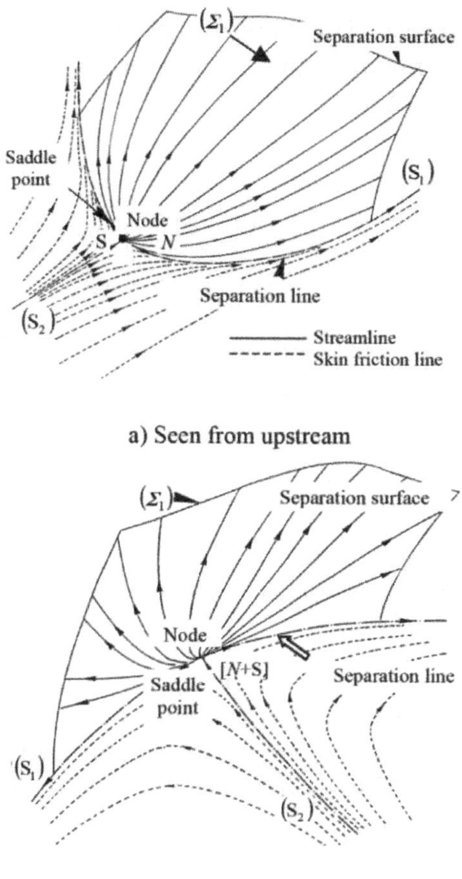

b) Seen from downstream

Figure 2.7. *Separation surface of the separation (detachment) type*

Let us consider the streamsurface (Σ_2), the trace of which on the body is the separation line (S_2). This surface has in common with (Σ_1) the streamline (S_2). As shown in

Figure 2.8, (S_2) is a separation line for the streamlines constituting (Σ_1). This separation line originates from the node N, which is a half-saddle point in the surface (Σ_2). The streamlines of (Σ_2) located behind (Σ_2) are deflected when approaching (S_2) in such a way that the flow turns back to take the direction of the flow on the other side of (Σ_1). This property is true for any flow coming from downstream when it arrives close to (Σ_1).

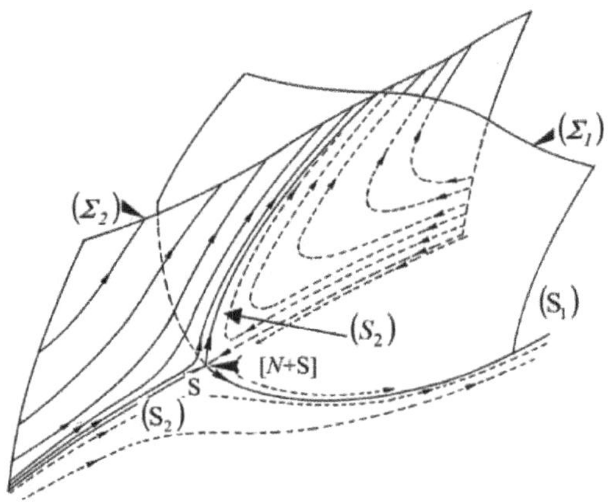

Figure 2.8. *Second separation surface and intersection of surfaces*

As already explained, when approaching the separation line, vorticity migrates from the thin boundary layer toward the outer flow. Then, convected by the streamlines, vorticity concentrates near the separation surface. At large distances, the vorticity layer surrounding the separation surface can be seen as a surface (Σ') supporting a distribution of vortices (in the mathematical sense), the flow being irrotational outside of (Σ'). In this scheme, (Σ') is a discontinuity surface called a *vortex sheet*. This kind of modeling is used to calculate separated flows as an extension of the method of singularities (or panel method). In reality, (Σ') is a mathematical concept distinct from the surface (Σ), which is a streamsurface of the real flow. Even if (Σ') is close to (Σ)

(which justifies the mathematical model), no confusion has to be made between the two surfaces. The surface (Σ) only has a physical signification and vorticity is continuously distributed in space (not concentrated on a surface). The set $[N + S + (\Sigma_1) + (\Sigma_2)]$ defines the flow structure – or topology – at separation.

The same description can be made – *mutatis mutandis* – in the vicinity of an attachment. As shown in Figure 2.9, an attachment node is associated with the saddle point S on the wall, where the streamlines constituting the attachment surface (Σ_1) terminate.

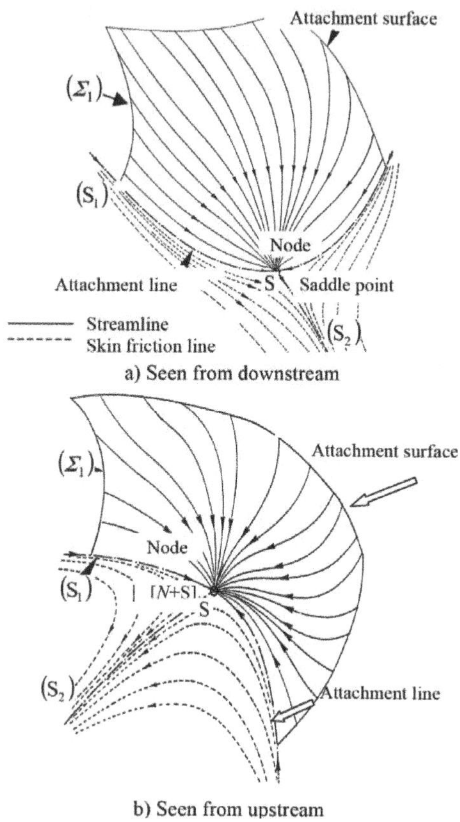

Figure 2.9. *Separation surface of the attachment type*

2.4. Vortical structures

In some circumstances, the separation surface (Σ_1) wraps up as shown in Figure 2.10, its trace (S_1) on the body surface rolling up around a critical point F of the focus type. Other skin friction lines roll up around F and "disappear" in F that is the trace on the body surface of a line (F) forming the axis of a three-dimensional critical point (see Figure 1.1). The external streamlines take a helical shape, the flow swirling around F. As shown in Figure 2.10(b), the pseudo-streamlines of the field projected in a downstream plane (P) roll up around a focus which is the intersection of (P) with the line (F). Such a flow is called a *vortical structure* or more briefly (but improperly) a vortex. In the present case, where the structure is in contact with a material surface, it is called a *tornado-like vortex*.

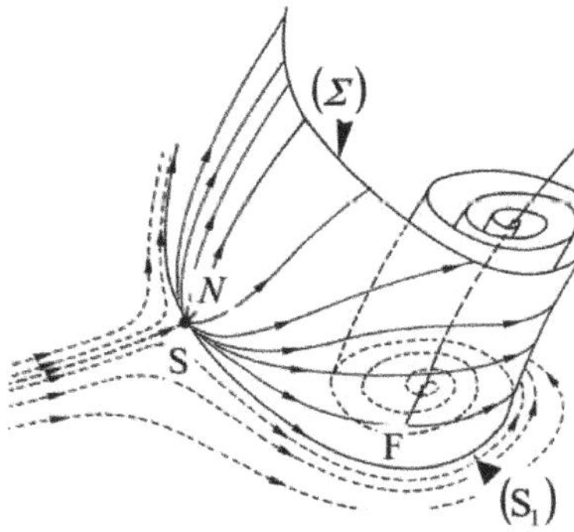

a) Wrapping of the separation surface

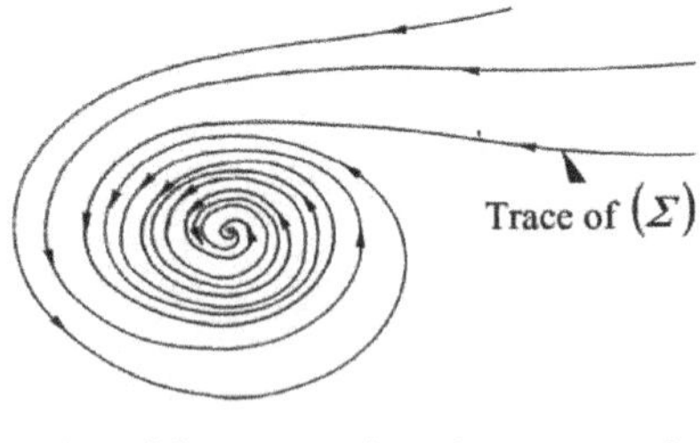

b) Section of the structure by a downstream plane

Figure 2.10. *Formation of a tornado-like vortical structure*

In other circumstances, the separation line (S_1), intersection of the streamsurfaces (Σ_1) and (Σ_2) of Figure 2.8, rolls up into a focus F, as shown in Figure 2.11. Then, the separation surface (Σ) wraps up to constitute a *horseshoe vortex*. The section of this structure by a plane is sketched in Figure 2.11(b).

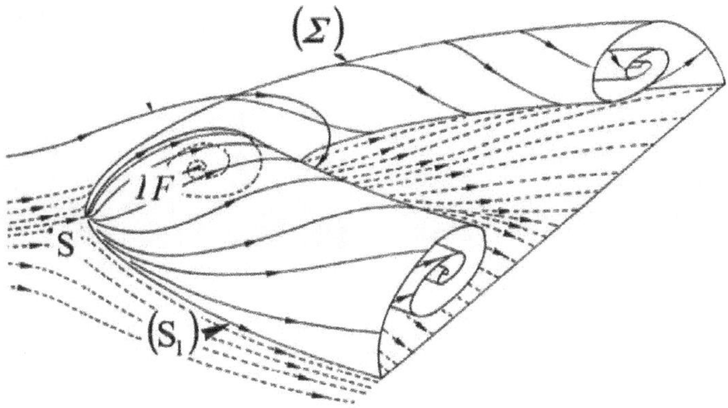

a) Wrapping of the separation surface

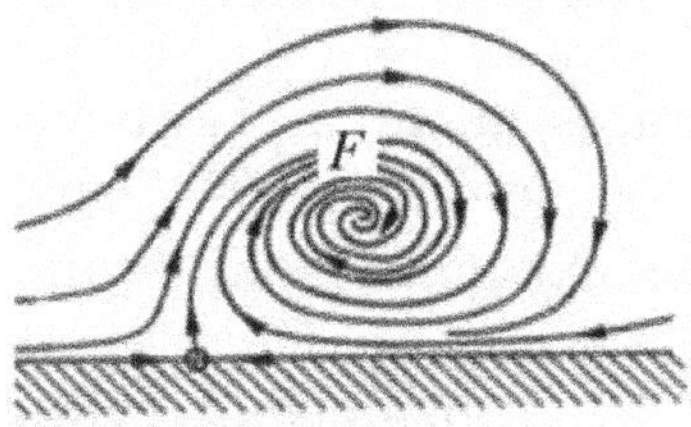

b) Section of the structure by a downstream plane

Figure 2.11. *Formation of a horseshoe vortex*

As seen above, during a separation process, the vorticity initially contained in a thin boundary layer leaves the wall region to concentrate in the vicinity of the separation surface. The rolling up of this surface leads to a spatial concentration of the vorticity generated in the boundary layers. Thus, the vortex contains a dissipative core around its axis (F), which is characterized by a peak of vorticity. A deficit of stagnation pressure corresponds to this peak. Figure 2.12 shows examples of a tornado-like vortex and a horseshoe vortex.

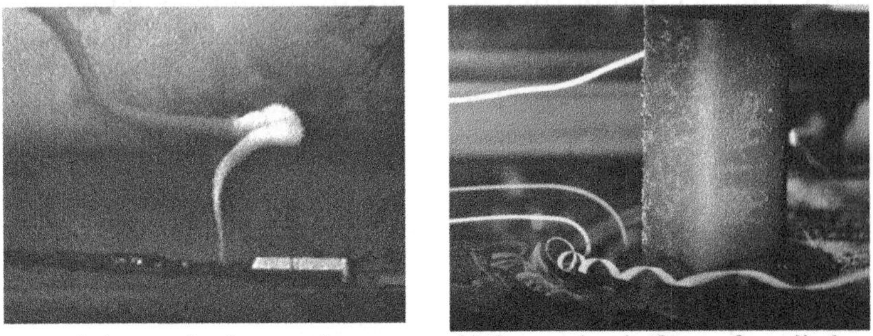

a) Atmospheric tornado b) Separation in front of a cylinder

Figure 2.12. *Examples of tornado-like vortex and horseshoe vortex formation*

2.5. Some properties of a vortical structure

Frequently, vortices adopt a regular structure around a nearly rectilinear pseudo-axis (F), such as the vortices emanating from the extremities of an aircraft wing or those forming over a delta wing (see Chapter 4). The velocity field is then defined by considering cylindrical coordinates $(r,\ \theta,\ z)$, where z is the distance along (F), r is the radial distance and θ is the angular coordinate (see Figure 2.13). As the structure is nearly axisymmetric around (F), the velocity field is represented by the following components:

– axial velocity $V_z(r)$ along z;

– tangential (or swirl) velocity $V_\theta(r)$;

– radial velocity $V_r(r)$ along r.

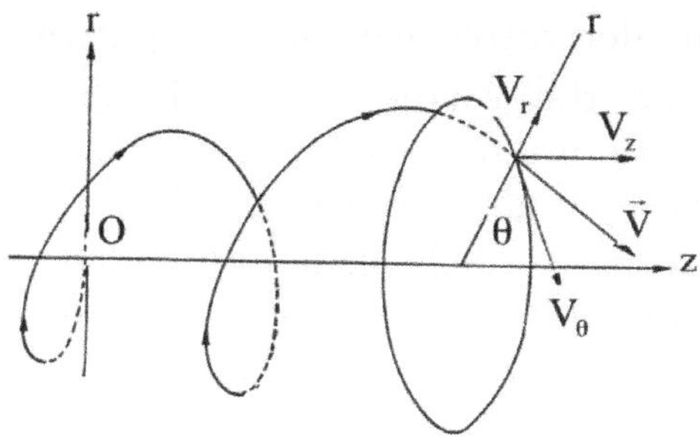

Figure 2.13. *Representation of an axisymmetric vortical structure*

In a well-established vortex, the radial component $V_r(r)$ can be frequently neglected. Different analytical expressions have been proposed to represent the velocity distributions in an axisymmetric vortex. These relations are either purely empirical or exact solutions of the motion equations. The following relations are frequently used:

$$V_\theta(r) = \frac{\Gamma_0}{2\pi r}\left\{1 - \exp\left[-1.256\left(\frac{r}{r_c}\right)^2\right]\right\}$$

$$V_z(r) = V_{z(ext)} + \left[V_{z(axis)} - V_{z(ext)}\right]\exp\left[-\left(\frac{r}{b}\right)^2\right]$$

where r_c is the viscous core radius (radial distance of the maximum swirl velocity), $V_{z(ext)}$ is the axial velocity outside the vortex, $V_{z(axe)}$ is the axial component along the vortex axis and b is a conventional thickness of the axial velocity profile defined by:

$$\frac{V_z - V_{z(axe)}}{V_{z(ext)} - V_{z(axe)}} = \frac{1}{e} = 0.368$$

From the radial equation:

$$\frac{\partial p}{\partial r} = -\frac{\rho V_\theta^2}{r}$$

we see that the derivative $\partial p / \partial r$ is negative. Consequently, the pressure on the vortex axis is lower than the outer pressure, the depression being more important as the rotation speed is higher. Because of this fact, in intense vortices (high swirl velocity), the axial motion is so accelerated that the component $V_{z(axis)}$ on the axis is greater than the outer value $V_{z(ext)}$.

The structure of a vortical structure, defined by the above relations, is illustrated in Figure 2.14. The axial profile $V_z(r)$ can either present a deficit on the vortex axis (wake-like profile) or a maximum (jet-like profile), the profile being flat in the intermediate case (see Figure 2.14(a)). Wake-like profiles (with a small deficit) are encountered in wing tip vortices, whereas jet-like profiles are typical of the very intense vortices forming over delta wings. The tangential (or swirl) profile $V_\theta(r)$ is shown in Figure 2.14(b). In the outer part of the structure, the profile follows a law of the form $V_\theta \approx 1/r$ corresponding to an irrotational vortex. In the axis region, where vorticity is concentrated, velocity is of the form $V_\theta \approx r$, which represents a solid body rotation: this part is the viscous core of the structure. The velocity is zero on the vortex axis $(r = 0)$. A transition zone ensures a continuous evolution of the velocity between the two regions. The swirling motion is also characterized by a circulation profile $\Gamma = 2\pi r V_\theta$, the factor 2π being frequently omitted. As shown in Figure 2.14(c), the circulation profile has a parabolic shape in the axis region and tends to a constant level Γ_0 outside the vortex where the flow is irrotational. At large distance, the vortex can be seen as a vortex filament, of intensity Γ_0, inducing in a point located at the distance r from the axis the swirl velocity $V_\theta = \Gamma_0 / (2\pi r)$. This schematization is used by the method of singularities or by the lifting line theory. In reality, the vortical structure is a continuous spatial distribution of vorticity. The static and stagnation pressure distributions corresponding to the profile of Figure 2.14(b) are shown in Figure 2.14(d): the rapid swirling motion induces a depression in the vortex's central part that is also a region of stagnation pressure deficit (or concentration of entropy). The temperature being lower in a vortex center, the vortices emitted during taking off by the wing tips and control surfaces of a plane are made visible by condensation of the atmospheric water vapor (if the weather is sufficiently humid).

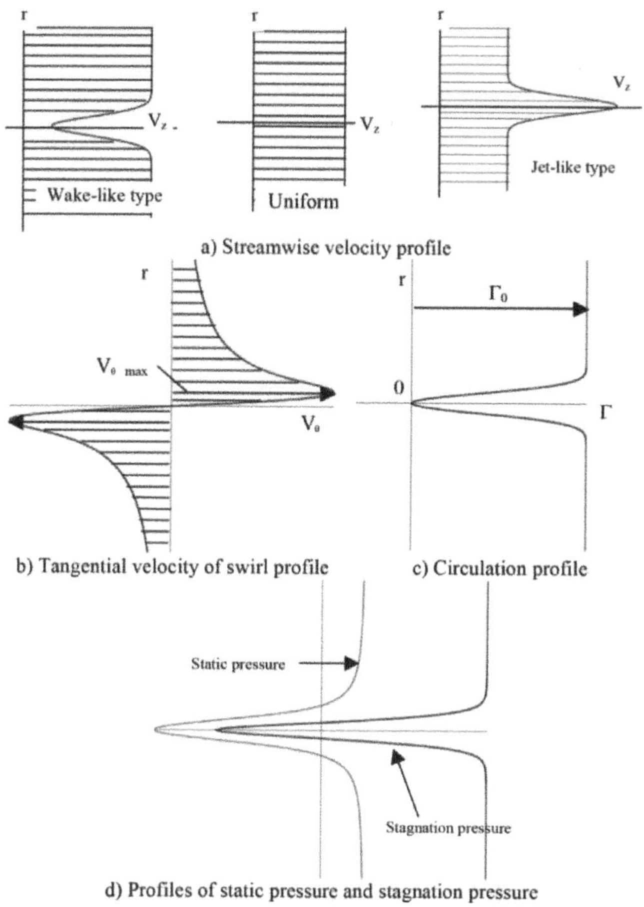

Figure 2.14. *Structure of an axisymmetric vortical structure.*

NOTE.– The term "vortex", used in place of vortical structure, can induce confusion with the notion of a vortex that designates the rotational of the velocity (curl of the velocity vector), which is a local property. Confusion can also be made with the mathematical entity, solution of the Laplace equation. In most cases, the context makes it possible to avoid confusion. In what follows, for the sake of simplicity, the denomination *vortex* will be used to designate what is in reality a vortical structure. Now the tendency is to call vorticity the module of the local velocity rotational.

3

Separated Flow on a Body

The concepts defined in the previous chapters are used to give a rational definition of a separated flow on a body. Separation is linked to the presence of saddle points in the skin-friction line pattern, associated with particular nodes or foci. Two kinds of separation are defined, one giving rise to the horseshoe vortex, the other to the tornado like vortex. The case of the ring vortex is also considered.

3.1. Basic rules and definitions

For the majority of analyzed configurations, it is not possible to respect the proportion of reality, some flow regions having to be enlarged to make the critical points visible (nose of a delta wing or the tip of a fuselage, for example). This entails a deformation of the body, and hence of the velocity field, in order to make visible details whose consideration is necessary for the construction of the whole flow field. When undertaking this operation, care must be taken not to modify the field topology, that is the number and relative position of the critical points, as well as of the separation lines. We have to perform a progressive deformation of the object by elongating some of its parts and contracting some others. The field picture thus obtained is topologically equivalent to that of the real flow. The sense of the skin friction lines or of the streamlines must also be respected. This fact can be intuitively understood without going into a mathematical definition.

We must rely on the surface flow pattern in order to detect the separation lines and to identify the critical points through which these lines pass. This examination is frequently difficult due to the lack of sharpness of the surface flow visualizations or the numerical calculation resolution. Frequently, the entire surface flow pattern is not visible because the represented field is too small, or because the whole pattern cannot be isolated (in the case of a model held by a support in a wind tunnel). In this case, we have to imagine a part of the surface flow pattern or fictitiously extend it in the upstream or downstream direction. This operation must be done by using the principle that the skin friction lines must originate at one or several attachment nodes and terminate in either a focus or a detachment node. A skin friction line cannot originate or end "somewhere"; we have to follow its path toward the terminal focus or node. Two skin friction lines cannot cross, except at a critical point. Any construction that would lead to the crossing of two skin friction lines elsewhere than at a critical point is incorrect. In the establishment of the skin friction line pattern, formula [3.2] from section 3.2 is of limited help. We have to rely on elementary topological rules forbidding the construction of "impossible" objects. Most of the time, formula [3.2] is applied at the end of the process as a checking step.

To continue the construction, it is mandatory to consider the separation and attachment surfaces belonging to the outer field. At the same time, we have to inspect the field in symmetry planes (if any) or/and in various projection planes, ensuring the topological consistency of the different pictures thus obtained and the field continuity at their junctions.

The separation surfaces' wrapping motion to form vortical structures must be carefully represented by considering the sense of the flow along the streamlines. Each separation surface must be linked to a separation line on the body and each vortex must be linked to a focus that can be on the body surface (for a tornado-like vortex) or in a plane of the flow (for a horseshoe vortex). A vortex cannot appear or end inside a flow field: its extremities must be in contact with a material surface, an interface (free air–water surface), or the vortex must close on itself (toroidal or ring vortex). The horseshoe vortex is not an exception to the rule: its two branches develop at a large distance to join with themselves forming a ring vortex. Such conditions are formalized by the Helmholtz theorems that apply, in fact, to a non-viscous flow. In reality, viscosity progressively dissipates the vortices (which is a good thing), this destruction being very slow (except if a vortex breakdown occurs) so that vortices may survive well downstream of the studied field.

The different considered lines (skin friction lines, streamlines and pseudo-streamlines) must be oriented in order to show the direction of the motion. Lack of this information can lead to constructions that are topologically inconsistent or physically impossible. Sometimes it is necessary to examine surface flow visualizations directly in the wind tunnel in order to detect the flow direction or the rotation sense around foci. Of course, video recording is helpful but the lack of resolution may blur fine details. When building a separation surface (of the attachment or separation type), it is mandatory to trace its streamlines and locate the node at their origin, which must coincide with a saddle point of the skin friction line pattern, as said above.

3.2. General definition: the basic separated structures

To introduce a general and rational definition of a separated (or detached) flow, let us consider a simply connected body (without a hole or handles) having the shape of a large ellipsoid (see Figure 3.1). This body is set at an incidence α in a flow that is uniform at upstream infinity.

The simplest situation that can be conceived for its skin friction line pattern is shown in Figure 3.1(a). A node N_1 is located at the body nose from which all the skin friction lines enveloping the body emanate. These lines terminate at a separation node N_2 at the rear extremity of the body. The flow streams past the body without containing any particular surface, except the body surface.

Starting from the previous situation, let us progressively raise the incidence α. The surface flow pattern undergoes a deformation so that a situation is reached where a saddle point S_1 appears in the pattern (see Figure 3.1(b)). Two separation lines (separatrices) go through this saddle point constituting barriers in the surface flow pattern. One of them, called the separation line (S_1), plays a special role as it forbids some of the skin friction lines coming from N_1 to reach some regions of the body surface. To build a topologically consistent pattern, a third node N_3 must be introduced in order to "feed" the region located behind (S_1). Hence, a flow over a body is called separated if its surface flow pattern contains more than two nodes. Consequently, any three-dimensional separated flow contains at least one saddle point through which a separation line passes; hence, the statement:

A flow is separated (detached) if its surface flow pattern contains at least one saddle point.

In the configuration shown in Figure 3.1(b), the skin friction lines issuing from nodes N_1 and N_3 terminate in the separation node N_2. It is possible to construct the arrangement represented in Figure 3.1(c) where two foci F_1 and F_2 appear on the body surface. A part of the skin friction lines coming from N_1, as also the separation line (S_1), disappears into F_1 and F_2. To ensure the consistency of the surface flow pattern, a second saddle point S_2 must be introduced along with a second separation line (S_2) that separates the skin friction lines going into F_1 and F_2 from those terminating in N_2.

The three previous situations are the simplest, which can be conceived; however, a quasi-infinite number of separated configurations can be imagined with increasing complexity (that is containing more and more critical points) by complicating the body shape, increasing its incidence or changing the Reynolds number. We intuitively understand that the number of critical points cannot be arbitrary if the surface flow pattern has to maintain its topological consistency.

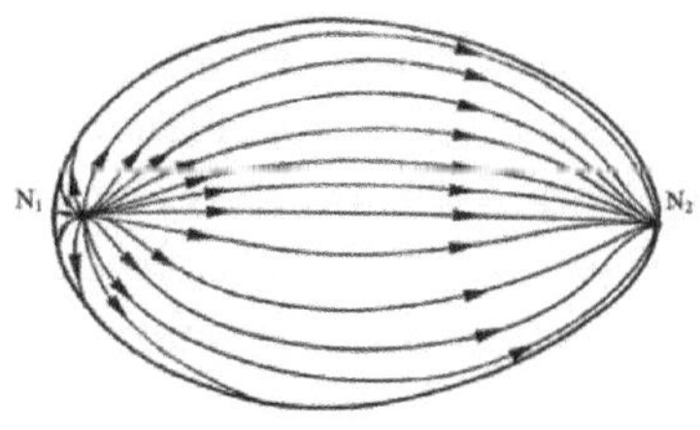

a) Non-separated flow

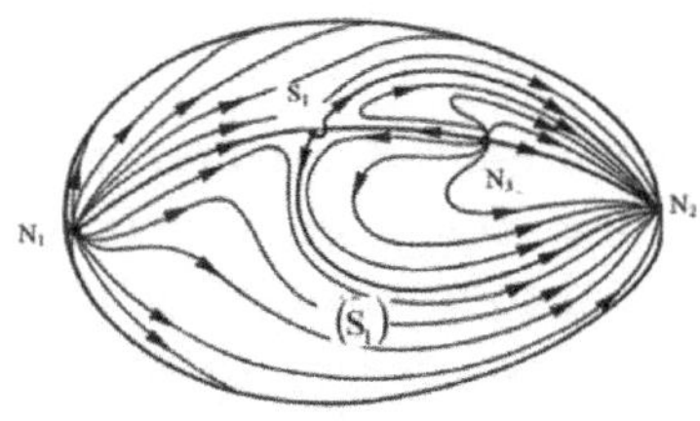

b) Separation with one saddle point and three nodes

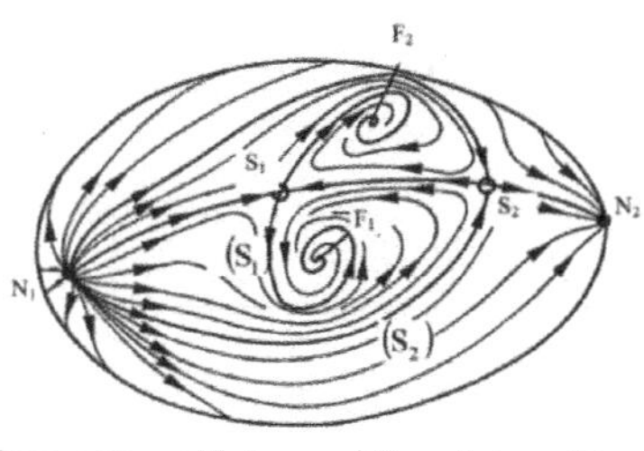

c) Separation with two saddle points and two foci

Figure 3.1. *Separation on a simply connected body Skin friction line pattern*

Nodes and foci are topologically equivalent, as they are sources or sinks for skin friction lines originating from them or ending in them. If we consider a complete and isolated body, the skin friction lines can only come from an attachment node. These lines must terminate at a point that can only be a node (or focus) of separation. If two attachment nodes are present on a body, a separation line must exist to demarcate the skin friction lines coming from each of these nodes, existence of a separation line entailing the presence of a saddle point. The same reasoning can be made with foci. On an isolated body, the number of nodes, foci and saddle points cannot be arbitrary. To be more precise, let us introduce the notion of complexity of a surface: a surface is said to be of complexity zero, or simply connected, if any closed curve traced on it can be reduced into a point by a continuous deformation without leaving the surface. This is the case of a sphere, an ellipsoid, a cylinder of finite length, a "potato", etc. Then, there is the following relation between the numbers of a critical type of different types on this surface:

$$\boxed{\sum (nodes + foci) - \sum saddle\ points = 2} \qquad [3.1]$$

For a non-simply connected surface of complexity different from zero, there exist closed curves traced on the surface that cannot be reduced to a point: this happens if the surface has a "hole" as a torus where a curve surrounding the hole cannot be continuously reduced to a point. If the surface has one hole, it is of complexity 1, if the surface has two holes, it is of complexity 2, etc. Then, relation [3.1] takes the form (Poincaré's formula):

$$\boxed{\sum (nodes + foci) - \sum saddle\ points = 2 - 2p} \qquad [3.2]$$

where p is the surface complexity.

Simple applications of the above formulas are illustrated in Figure 3.2. The first case (see Figure 3.2(a)) is a body having the shape of a camembert box, without a hole, hence it is limited by a surface of complexity zero. Let us draw a fictitious – but possible – surface flow pattern containing a node N_1 at the front part of the box, a separation with two foci F_1 and F_2, two saddle points S_1 and S_2 and a separation node N_2 at the rear (not visible on the picture). The upper and lower parts of the box being assumed symmetrical, there are two foci and two saddle points on the other side of the box, hence the following numbering:

$$\sum(nodes + foci) - \sum saddle\ points = 6 - 4 = 2 - 2 \times 0 = 2$$

Now the box has a hole as shown in Figure 3.2(b), the surface being of complexity 1. A separation line of the type shown in Figure 3.1(c) is present downstream of the hole, the patterns on the upper and lower parts still being symmetrical. The hole establishing a communication between the upper and lower sides, a separation line going through the saddle points C_3 and C_4 and located on the hole side (see Figure 3.2(b)) must exist. In this case, we have:

$$\sum(nodes + foci) - \sum saddle - points = 6 - 6 = 2 - 2 \times 1 = 0$$

These examples are among the simplest that can be devised. To have some fun, we can imagine bodies with 2, 3, ..., *n* holes, construct topologically consistent surface flow patterns and check the above formula.

Application of the above formulas on a model tested in a wind tunnel is rarely possible since the entire surface must be considered, which includes not only the model, but also its support and the wind tunnel itself. Its application to a limited portion of a surface flow pattern must be avoided. However, if the upstream and downstream conditions are well defined and if the flow is nearly uniform on the boundaries of the considered domain, then it is possible to apply the above relations by taking into consideration a node "at upstream infinity" – origin of the skin friction lines – and a node "at downstream infinity" – where the lines leaving the domain terminate.

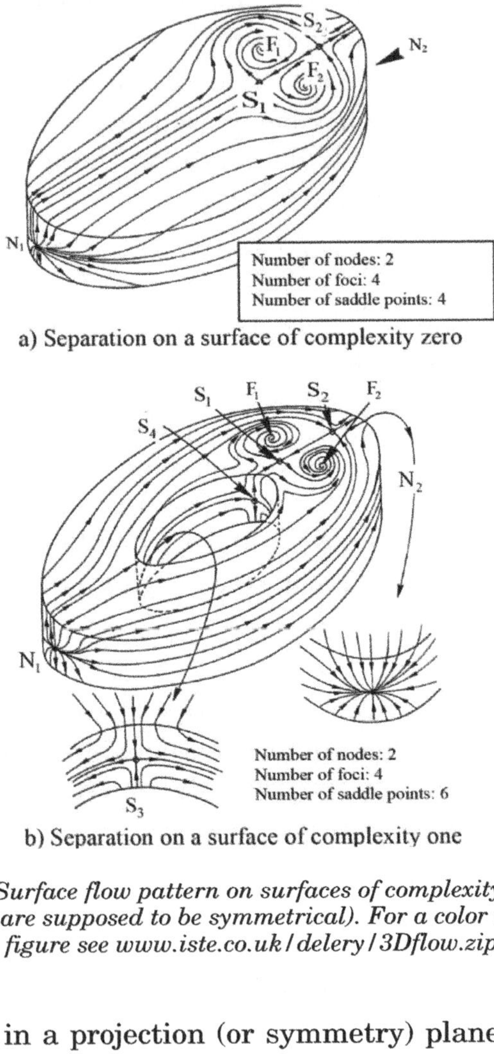

Figure 3.2. *Surface flow pattern on surfaces of complexity zero and one (the patterns are supposed to be symmetrical). For a color version of this figure see www.iste.co.uk/delery/3Dflow.zip*

The flow in a projection (or symmetry) plane can contain projected (or real) streamlines running through nodes or saddle points on the body surface. As the body occults a part of the flow, then we have to consider half-nodes or half-saddle points (see Figure 3.3). The relation linking the various critical points of the projected field then becomes:

$$\left[\sum(\textit{nodes}+\textit{foci})+\frac{1}{2}\sum\textit{half nodes}\right] - \left[\sum\textit{saddle points}+\frac{1}{2}\sum\textit{half saddle points}\right]=-1 \qquad [3.3]$$

Figure 3.3. *Field in a transverse section plane. Half-nodes and half-saddle points*

Physical grasping of a separated three-dimensional field necessitates going further in identification and numbering of the critical points observed on the surface. We have to complete the construction of the surface pattern by that of the outer field. The connection with the separation surfaces associated with the separation and attachment lines of this pattern must be ensured. This exercise is frequently arduous since it entails both imagination (we rarely have the complete information on the field), logical rigor (making an "impossible" object must be avoided) and the capacity to visualize objects in three dimensions. To a surface flow pattern, it is possible to associate several external fields, the choice of the solution being imposed by information provided by the experiment or the calculation. If this information is lacking, common sense leads us to adopting the simplest possible field organization by ruling out improbable or manifestly instable situations. Construction of the main components of the flow must be established as far as possible (in general, it is impossible to take into consideration all the details). Then, an encounter of a topological impossibility can undermine the entire construction. The field representation must be based on accurate drawings, as too crude sketches nearly always contain inconsistencies. The examples that follow will illustrate the method.

3.3. Field associated with a separation with one saddle point and three nodes: the horseshoe vortex

To simplify, we admit that the body and the flow have a symmetry plane containing the upstream velocity vector $\vec{V}_\infty$, which is frequently the case in reality. Let us consider the surface flow pattern with three nodes and one saddle point shown in Figure 3.1(b) connected to the basic separation found in nearly all separated flows.

Like the other skin friction lines, the separation line (S_1) terminates at the detachment node N_2. In the first situation, the separation surface (Σ_1) sustained by (S_1) winds as shown in Figure 3.4 to constitute a toroidal vortex. The trace of this vortex in the symmetry plane is made up of two "stable" foci F_1 and F_2, the streamlines winding around them. The two separation lines (S_1) and (S_2), respectively, issued from S_1 and N_2 wind around these foci. A third separation line (S_3), terminating at node N_3, separates the streamlines into two families: one family disappears in F_1 and the other family disappears in F_2. A forth separation line must be introduced to demarcate the streamlines going into the foci from those flowing downstream. The field consistency imposes a fifth separation line going through the saddle point located in the symmetry plane inside the flow field. One of its branches winds around F_2 and the other goes downstream.

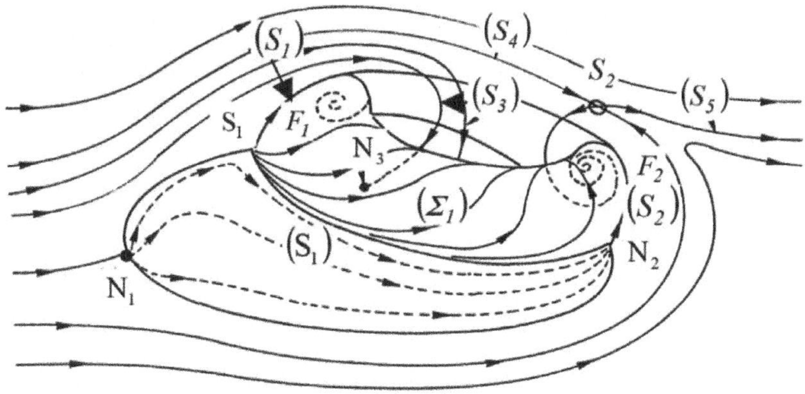

Figure 3.4. *Separation with one saddle point and three nodes. First construction. Separation surface*

Figure 3.5 gives an enlarged picture of the situation near the points S_1 and N_2. The streamlines constituting the separation surface (Σ_1) are issued from the node N_1 coincident with S_1. They are organized in two families separated by the separation lines (S_1) and (S_2). The downstream node N_2 coincides with the saddle point S_2 contained in (Σ_1). If we add a node at upstream infinity, origin of the streamlines, and another node downstream, where the streamlines terminate, the construction in Figure 3.4 is *complete* and topologically possible.

We are confronted with a difficulty to terminate the streamlines of the separation surface (Σ_1). As the focus F_2 is a winding focus, evacuation of these lines is problematic. The same difficulty exists for the streamlines close to (Σ_1). The construction of Figure 3.4 must be instable. The organization of the toroidal vortex can be modified by considering the flow represented in Figure 3.6. In this case, the surface pattern contains the saddle point S_2, in addition to N_1, N_2 and S_1. The separation line (S_1) going through S_1 and S_2 encloses skin friction lines organized in two symmetrical centers.

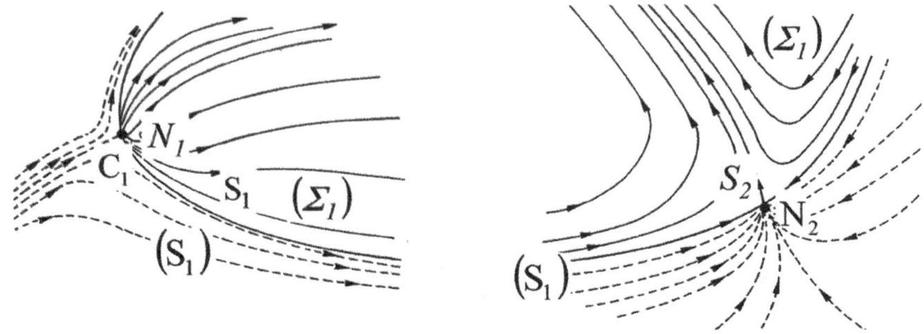

a) Vicinity of the saddle point S_1 b) Vicinity of the saddle point S_2

Figure 3.5. *Separation with one saddle point and three nodes. First construction. Details*

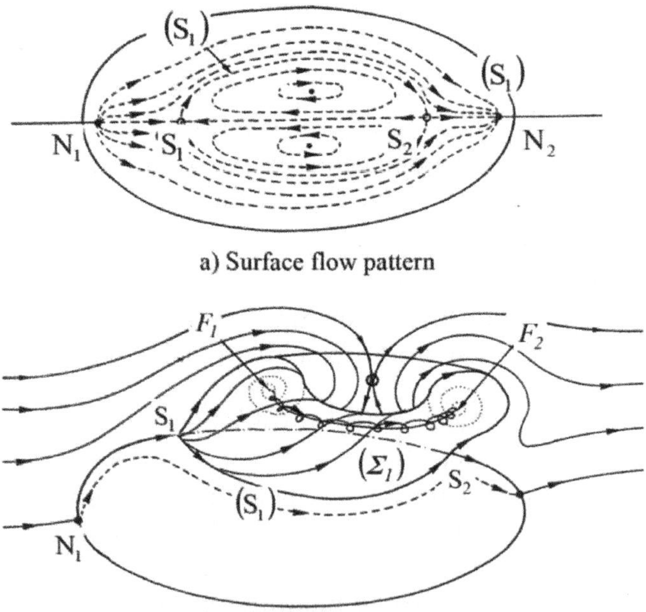

a) Surface flow pattern

b) Separation surface

Figure 3.6. *Separation with two saddle points and two nodes. Second construction*

The flow in the symmetry plane (see Figure 3.7) contains a winding focus F_1 and an unwinding focus F_2. The streamlines coming from F_2 are split in a family streaming downstream and a family that disappears into F_1. Similarly, the streamlines coming from upstream partly flow downstream and partly wind around F_1. Such circumstances impose the existence of a saddle point S_1 in the flow field. In this system, there is an exchange of fluid between foci, the streamlines coming from F_2 being evacuated in F_1. The fluid concentrated in the core of F_1 flows along the structure axis, as shown in Figure 3.6, and escapes through F_2. A separation node, the origin of the streamlines forming (Σ_1), is associated to S_1 and an attachment node to which streamlines converge, corresponds to the saddle point S_2. A stable state can probably exist; however, the existence of two centers is improbable with this type of organization.

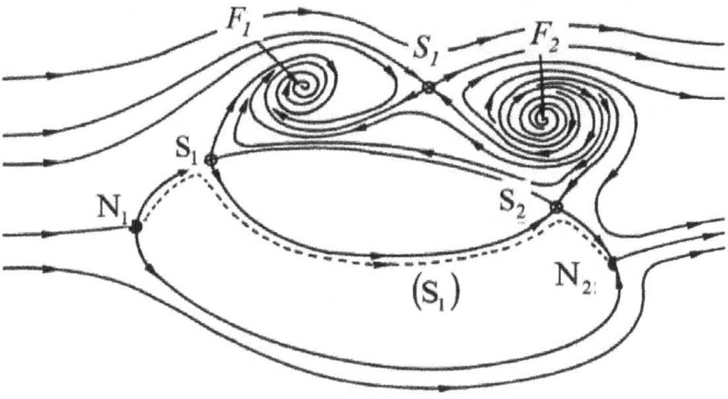

Figure 3.7. *Separation with two saddle points and two nodes. Second construction. Flow in the symmetry plane*

It is possible to conceive an unsteady situation in which the vortical structure detaches from the body surface in the form of a ring vortex carrying away the vorticity that has migrated in the vortex according to the mechanism explained above. This ring, convected by the flow, can be visualized as a ring of smoke like the ring ejected by the volcano Etna shown in Figure 3.8.

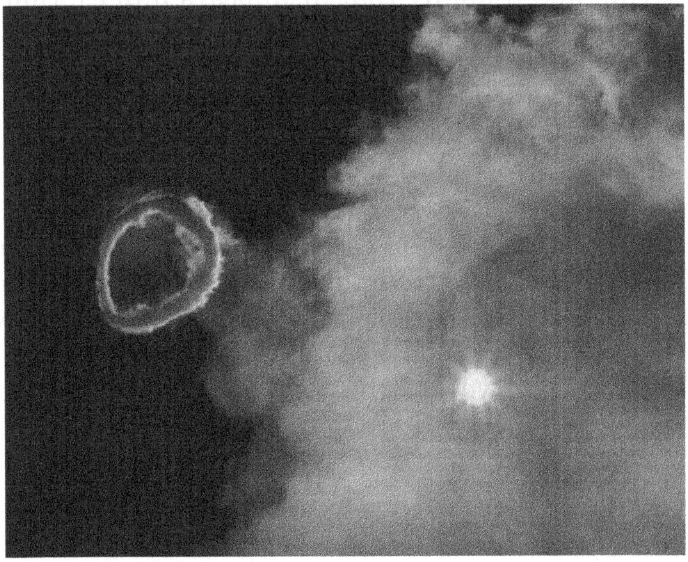

Figure 3.8. *Toroidal, or a smoke vortex, emitted by Etna volcano (Pour la Science, May 2006)*

When two centers coexist on the body surface, a toroidal vortex is the most simple to conceive as drawn in Figure 3.9. The separation line (S_1), going through the saddle points S_1 and S_2, is the trace of the separation surface (Σ_1), which has the form of a dome. The streamlines of (Σ_1) originate at the separation node N_1 and terminate at node N_2. Inside the volume delimited by (Σ_1), the streamlines are coplanar closed curves turning around the centers. The flow is organized in a closed separation bubble inside which the flow is trapped and involved in a recirculation motion. This organization, found in planar or in axisymmetric two-dimensional flows, is highly improbable considering its very special character (see Chapter 6). By adding nodes at the two infinities, the configuration is consistent and complete.

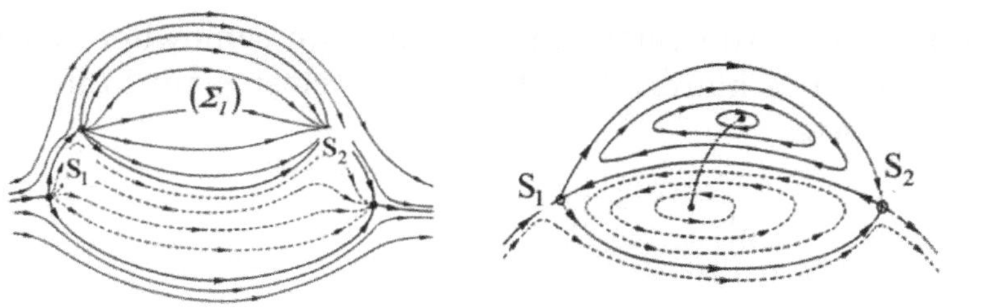

Figure 3.9. *Separation with two saddle points and two nodes. Second construction. Closed vortex in contact with the surface*

The preceding constructions being not satisfactory because of their either instable or improbable nature, a simpler organization has to be looked for. For this, let us consider the picture in Figure 3.10 whose surface flow pattern is the pattern of Figure 3.1(b). The separation surface (Σ_1) is still sustained by the separation line (S_1) passing through the saddle point S_1 and the node N_2. The streamlines belonging to (Σ_1) originate at the separation node N_1 in coincidence with S_1. This surface wraps to constitute the vortex that "escapes" in the downstream direction taking the shape of a horseshoe vortex. The intersection of this vortex with a plane sensibly perpendicular to the upstream velocity exhibits two foci corresponding to the two legs of the horseshoe vortex. On (Σ_1), the saddle point S_2 is associated with N_2; the separation line (S_1) going through S_2 splits the streamlines issuing from N_1 into two families, each set of lines winding around one of the vortex legs.

This organization is frequently observed downstream of a blunt body, for example an automobile. Downstream of the body, the two vortices (in reality, the two legs of the same vortex) constitute the vortex wake of the vehicle (in reality, other separations occur on the vehicle so that the wake is made up of several such vortices as seen in section 5.4). The flow represented in Figure 3.10 has the merit of being simple, which is a warranty of reality; however, it is not complete since the vortex must terminate somewhere either by ending on a material surface, or by closing up to make a toroidal vortex.

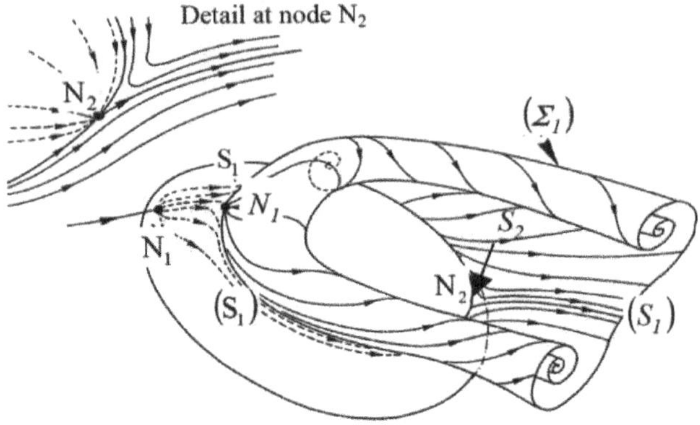

Figure 3.10. *Separation with one saddle point and three nodes. Formation of a horseshoe vortex. Separation surface*

3.4. Field associated with a separation with one saddle point and two foci: the tornado-like vortex

A frequently encountered surface flow pattern is the pattern shown in Figure 3.1(c) where two foci and two saddle points exist, in addition to the attachment and detachment nodes. To simplify the description of the flow, it is assumed that the configuration is symmetrical, the two saddle points being on the same separation line. This circumstance is improbable, but introduction of some asymmetry does not entail a complete change of the topology as seen below.

The elements associated with this pattern are:

– the separation line (S_1), going through the saddle point S_1 and spiralling around the foci F_1 and F_2, is the trace of a separation surface (Σ_1) that wraps to constitute the two tornado-like vortices whose traces on the body are the foci F_1 and F_2 (see Figure 3.11). The streamlines of (Σ_1) are issued from the node N_1 coincident with the saddle point S_1.

– the separation line (S_2) going through the saddle point S_2 has the character of an attachment line. It is the trace on the surface of the attachment surface (Σ_2), the streamlines of which terminate at the attachment node N_2 (see Figure 3.12).

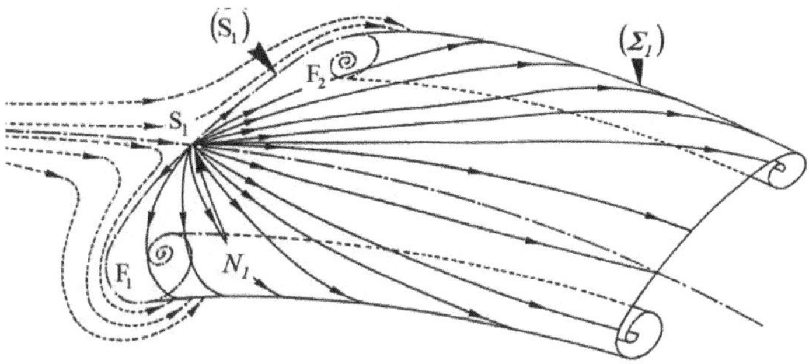

Figure 3.11. *Separation with one saddle point and two foci. Formation of tornado-like vortices. Separation surface*

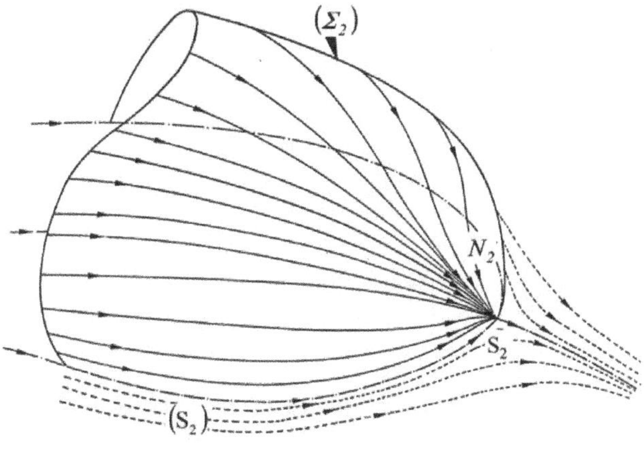

Figure 3.12. *Separation with one saddle point and two foci. Attachment surface For a color version of this figure see www.iste.co.uk/delery/3Dflow.zip*

Since the two tornado-like vortices have to escape downstream, the separation surface must be crossed by the separation surface. As seen in Figure 3.13, intersection of surfaces (Σ_1) and (Σ_2) gives in (Σ_2) a structure similar to that observed at the wall, the surface containing two foci F_1 and F_2. The streamlines disappearing in these foci are separated from those reaching the attachment node N_2 by the separation line going through the saddle point S_1. The separation line (S_1), which separates the streamlines coming from upstream from those spiraling around F_1 and F_2, passes through the saddle point S_1.

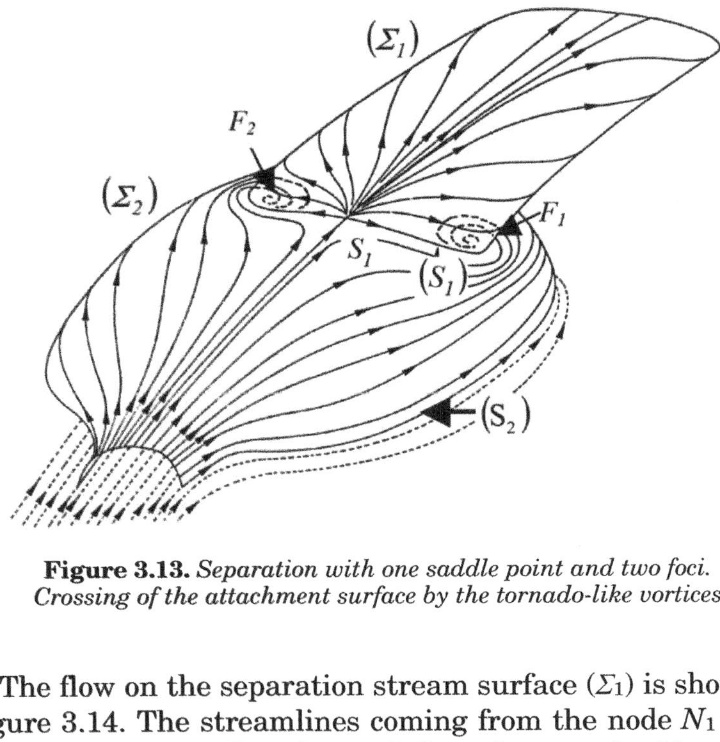

Figure 3.13. *Separation with one saddle point and two foci. Crossing of the attachment surface by the tornado-like vortices*

The flow on the separation stream surface (Σ_1) is shown in Figure 3.14. The streamlines coming from the node N_1 move away from the wall and are deflected on each side of the separation line going through the node S_3 when approaching the separation line (S_3), intersection of (Σ_1) and (Σ_2). The streamlines undergo a spiraling motion into the core of the tornado-like vortices. Above (S_3), the streamlines emanate from node N_3 that constitutes with S_3 the combination [node + saddle point] (N_3 + S_3). These streamlines are part of the separation surface (Σ_1), which have crossed the surface (Σ_2).

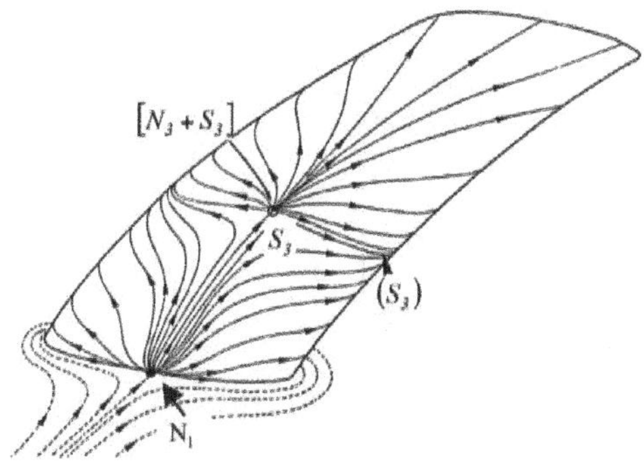

Figure 3.14. *Separation with one saddle point and two foci. Flow on the separation surface*

As shown in Figure 3.15, the streamlines forming the attachment surface come from upstream (the node at upstream infinity). On approaching the separation line (S_3), they are split into two families separated by the separation line going through the node S_3. Thereafter, they wind around the foci F_1 and F_2. Figure 3.16 shows the attachment surface seen from downstream.

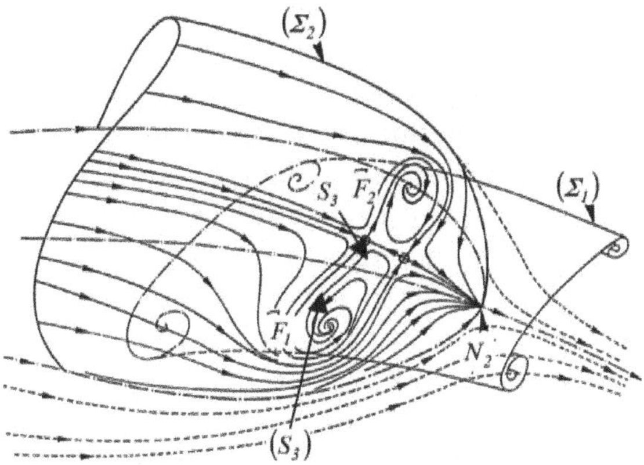

Figure 3.15. *Separation with one saddle point and two foci. Flow on the attachment surface*

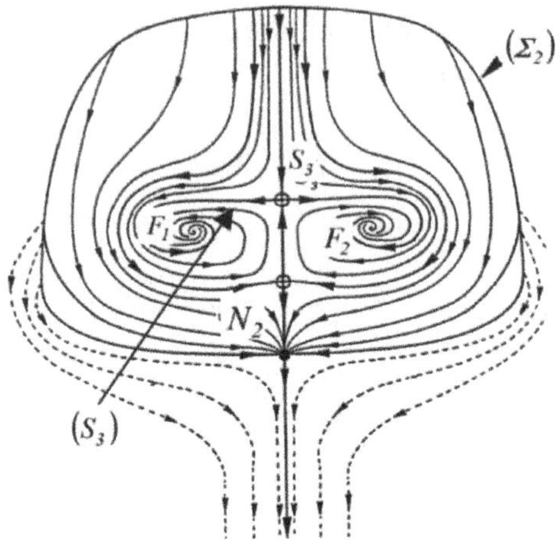

Figure 3.16. *Separation with one saddle point and two foci. Attachment surface seen from downstream*

The above structure is, in fact, incomplete. In addition, of the two nodes at the upstream and the downstream infinity, we have to imagine that the two tornado-like vortices linked to the foci F_1 and F_2 either end on a material surface, or constitute a ring vortex by closing in on themselves. We may object that the presence of two saddle points in the above construction located on the same separation line (as is the case for S_1 and S_2) is improbable, the real flows never being perfectly symmetrical. In addition, according to some authors, such a configuration would be unstable, which is another way to say the same thing. Thus, we will substitute for the flow pattern in Figure 3.17(a) the pattern drawn in Figure 3.17(b), where the two saddle points S_1 and S_2 are no longer on the same separation line. A third separation line (S_3) must be introduced to separate the skin friction lines disappearing in F_1 from those going into F_2. The attachment line (S_2) separating the skin friction lines going into the foci F_1 and F_2 from those flowing downstream still exists. In reality, for a two-dimensional test arrangement, the two patterns are very close to each other and only a trained eye can notice the surface pattern asymmetry. The construction of the separation surfaces linked to the asymmetric surface flow pattern can also be made at the price of a modest complication. In what follows, we will admit that the surface flow pattern has a certain symmetry, to avoid too much complication.

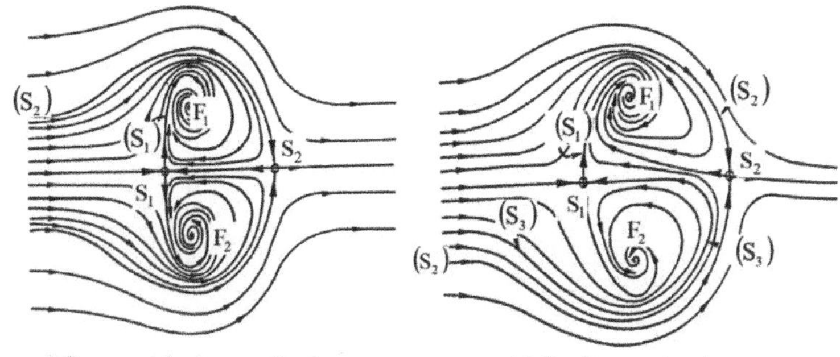

Figure 3.17. *Separation with two saddle points and two foci. Surface flow pattern variants*

4

Vortex Wake of Wings and Slender Bodies

The topology of the delta wing at incidence is analysed by considering the skin-friction line pattern and the vortex system forming over the wing, with explanation of the vortex lift. The similar analysis is applied to a slender body with a sharp apex. The case of the classical wing with moderate sweep angle is studied to show the mechanism leading to the formation of the tip vortices and wing wake.

4.1. Vortical structures over a delta wing

A delta wing at incidence constitutes an important case of a three-dimensional geometry because of its simple shape and practical importance (combat aircraft as well as supersonic transport aircraft are equipped with delta wings). Let us consider a delta wing with a sweep angle φ set without yaw at an incidence α in a uniform upstream flow. The wing leading edge is rounded and the apex is blunt. In some studies, a wing with a sharp leading edge and pointed apex is considered in order to simplify flow modeling because the main separation line can be intuitively fixed along the leading edge. However, for the physical understanding of the flow, it is easier to consider a rounded leading edge because of the singular character of a pointed or sharp object with zero thickness. In what follows, we will not consider phenomena occurring at the wing trailing edge where the flows coming from the upper and lower wing surfaces interact.

The wing incidence α is raised until the establishment of a separated situation with the occurrence of a saddle point. At the wing nose, the skin friction lines pattern (see Figure 4.1(a)) shows an attachment node N_1 feeding the streamlines of the wing lower surface. In the wing symmetry plane, there is a saddle point S_1 on the upper surface through which the separation line (S_1) running along the leading edge flows (see Figure 4.2(b)). On the wing's lower surface (see Figure 4.1(a)), the skin friction lines coming from N_1 are split into two families by the separation line (A_1) of the attachment type contained in the symmetry plane. These skin friction lines flow around the leading edge and converge toward the separation line (S_1). To feed the part of the surface pattern between (S_1) and the symmetry plane, a second attachment node N_2 exists behind the saddle point. In the plane of symmetry, a separation line (A_2), of the attachment type, splits the skin friction lines into two families flowing on each side of the wing. Convergence to the separation line (S_1) is very rapid from the wing's lower surface but more progressive on the wing's upper surface.

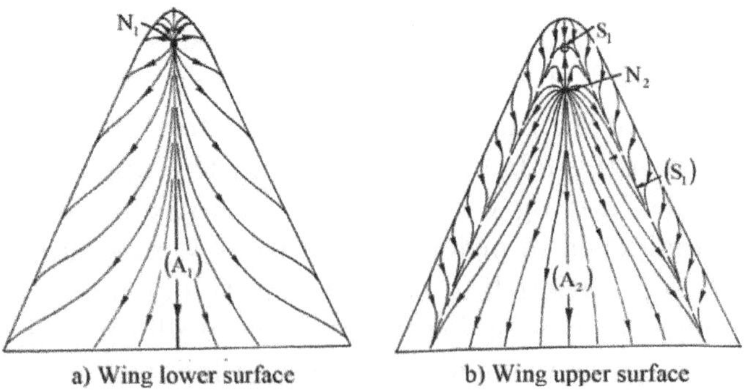

Figure 4.1. *Separation over a delta wing. Surface flow pattern*

To the separation line (S_1), a vortical structure is attached resulting from the wrapping of the separation surface (Σ_1) (see Figure 4.2). This structure has the shape of a horseshoe vortex with a focus F_1 in its symmetry plane, as shown in Figure 2.4. The streamlines of (Σ_1) are issued from the node N_1 coincident with S_1 on the surface. The saddle point S_1 is associated with node N_2 where the separation line (S_1), dividing the streamlines that spiral into F_1 from those that flow downstream, ends.

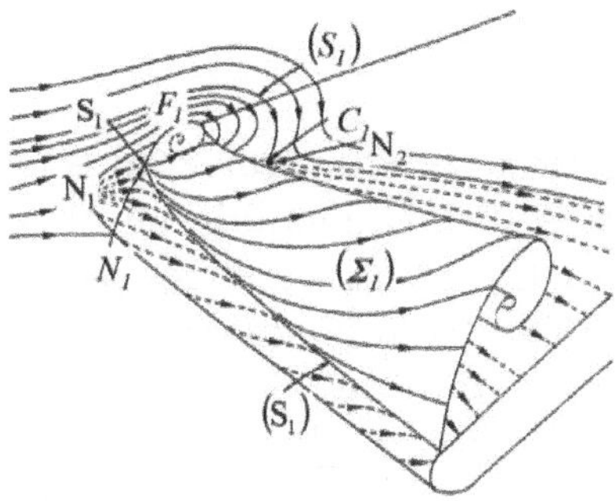

Figure 4.2. *Separation over a delta wing. Primary separation surface*

Figure 4.3(a) shows the organization of the pseudo-streamlines projected in a plane perpendicular to the wing's upper surface. These lines are oriented according to an ascending motion from a node at infinity "downstream". This organization contains two foci F_1 and F_2, which are the traces in the projection plane of the horseshoe vortex core. The pseudo-streamlines spiraling around F_1 and F_2 are separated from those flowing to "above" infinity by the separation line (S_3) going through the saddle point S_3. The structure made up of two foci and a saddle point in the symmetry plane is frequently observed in the separated flows. Figure 4.3(b) shows a water tunnel visualization of the flow above a delta wing. This picture is obtained by illuminating a plane normal to the wing's upper surface and photographing air bubbles used as tracers.

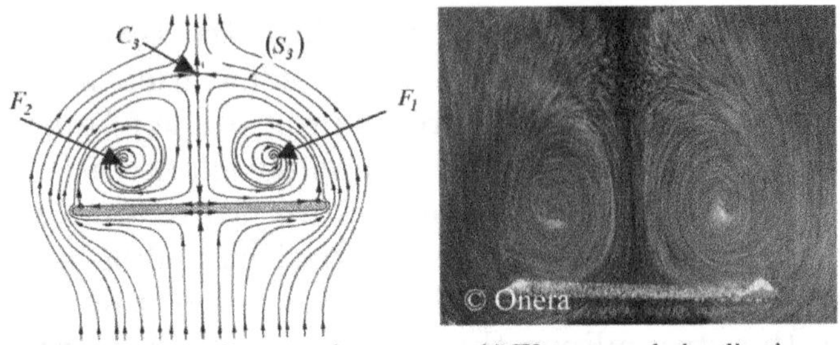

a) Pseudo-streamlines topology b) Water tunnel visualization

Figure 4.3. *Separation over a delta wing. Field projected in a plane perpendicular to the wing*

Other organizations of the projected flow are possible. In the case of Figure 4.4(a), the saddle point S_3 becomes a half-saddle point S_3 on the wing's surface where the separation line (S_3) ends up. The pseudo-streamlines located above (S_3) turn past the vortex and flow along the upper surface in the direction of the symmetry plane. Before reaching it, they suddenly bend to orient themselves upward to infinity "above". The half-saddle point S_3 must be located on an attachment line (A_2) belonging to the surface pattern. This line separates the skin friction lines flowing toward the separation line (S_1) from those slowly converging towards the symmetry plane (see Figure 4.4(b)).

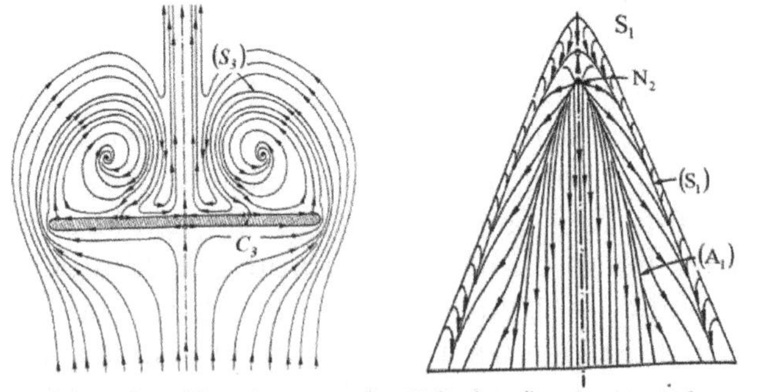

a) Field projected in a plane normal to the wing b) Surface flow pattern on the upper surface

Figure 4.4. *Separation over a delta wing. Other organization*

In most cases, the flow over a delta wing contains several vortices, and the flow under the vortex separates when it meets the flow coming from the lower surface. In the organization shown in Figure 4.5, the *primary vortex*, resulting from the separation along (S_1), coexists with a *secondary vortex* originating from the secondary separation line (S_2).

Then, the surface flow pattern contains two separation lines (S_1) and (S_2) (see Figure 4.6(a)). As each of these lines is a convergence line for the skin friction lines, a separation line of the attachment type (A_2) must exist between them. We admit without demonstration that an attachment line must exist between two separation lines for obvious reasons of topological consistency (and vice versa).

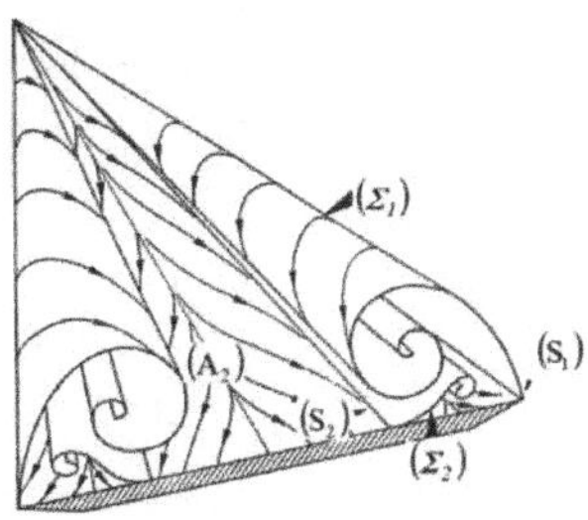

Figure 4.5. *Separation over a delta wing. System with two separation surfaces*

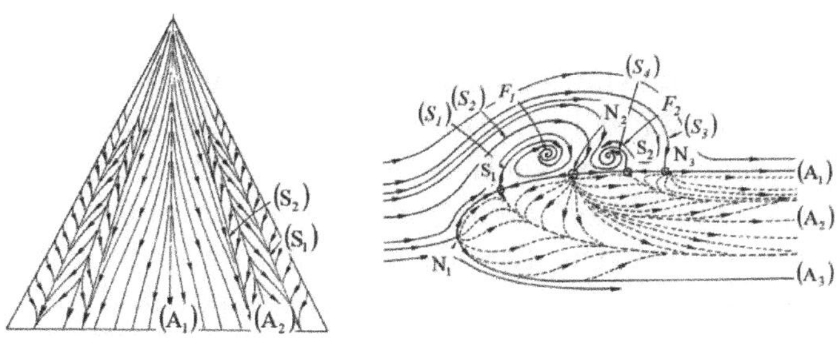

a) Surface flow pattern b) Surface flow pattern in the apex region

Figure 4.6. *Separation over a delta wing. System with two vortices*

In the apex region, the surface pattern includes three nodes (N_1, N_2 and N_3) and two saddle points (S_1 and S_2) in the symmetry plane (see Figure 4.6(b)). The separation lines (S_1) and (S_2) go through S_1 and S_2 whereas the attachment lines (A_1), (A_2) and (A_3) originate at the nodes N_1, N_2 and N_3, respectively. The traces of the vortical structures in the symmetry plane are the foci F_1 and F_2 corresponding to the primary and secondary vortices. The flow in the symmetry plane includes the four separation lines (S_1), (S_2), (S_3) and (S_4) (see Figure 4.6(b)). As shown in Figure 4.7, the pseudo-streamlines structure in the projection plane contains the foci at the intersection with the horseshoe vortices axes, the saddle point located on the symmetry axis, above the wing, and the half-saddles.

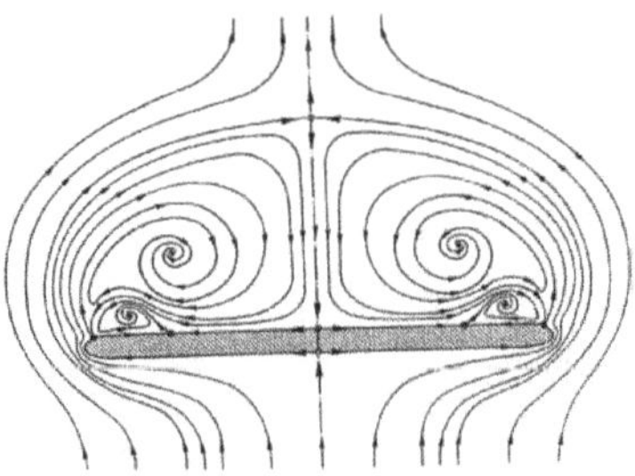

a) Field projected in a plane normal to the wing

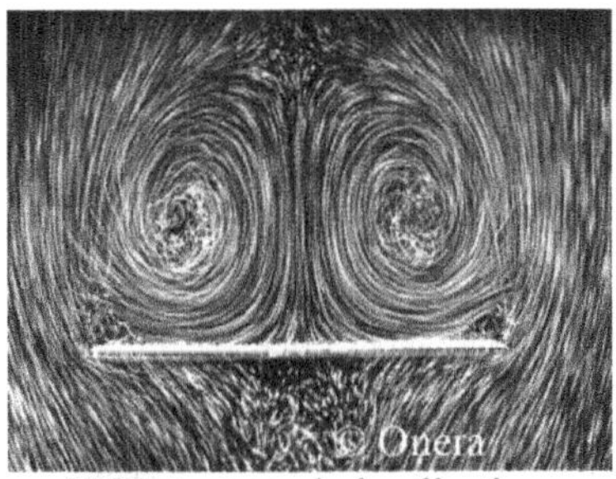

b) Water tunnel visualization

Figure 4.7. *Separation over a delta wing. The two-vortex system*

In the organization represented in Figure 4.8, the pseudo-streamlines unwind from the trace F_3 of the primary vortex (instable focus, see section 1.3.2.3). Since the pseudo-streamlines must have a converging motion in the outer part of the vortex core (the flow is from the lower surface to the upper surface), the two families are separated by a *limit circle* or *limit cycle* to which the pseudo-streamlines of each family tend asymptotically. This phenomenon is classical in the dynamic system theory, such cycles representing self-sustained oscillations. Although the reality of the phenomenon is well established in the present case, it must be considered with circumspection. In particular, we should avoid establishing a correlation between the existence of a limit circle (due to an unstable focus) and the vortex susceptibility to breakdown. We should not forget that the projection of the velocity field does not give an objective view of the real flow.

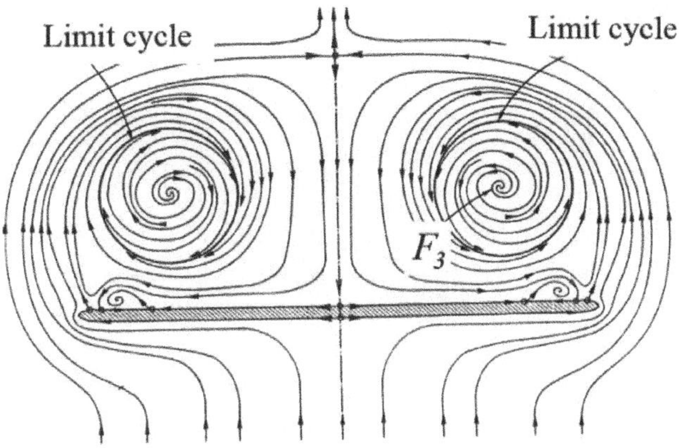

Figure 4.8. *Separation over a delta wing. Projected field with limit cycles.For a color version of this figure see www.iste.co.uk/delery/3Dflow.zip*

Delta wing flows with more than two vortices are observed, the existence of these different vortical regimes depending on several parameters. Thus, the number of vortices decreases when the Reynolds number increases. At a high Reynolds number, a two-vortex system is generally observed. The sweep angle, the angle of incidence and the wing plane form also play a role. Figure 4.9 is a water tunnel visualization of the flow past, from a model of the Concorde supersonic transport aircraft. The filaments of dye emitted near the wing's apex visualize the vortices forming above the wing's upper surface. A bubble visualization of the flow in a transverse plane is shown in Figure 4.10 where the two big vortices are clearly visible.

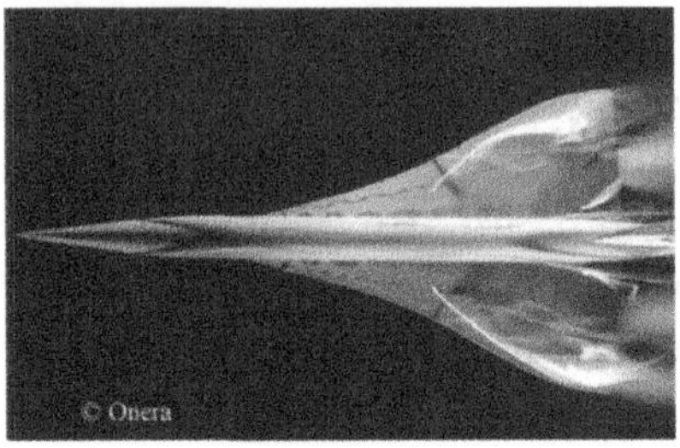

Figure 4.9. *Water tunnel visualization of the vortices over the wing of Concorde*

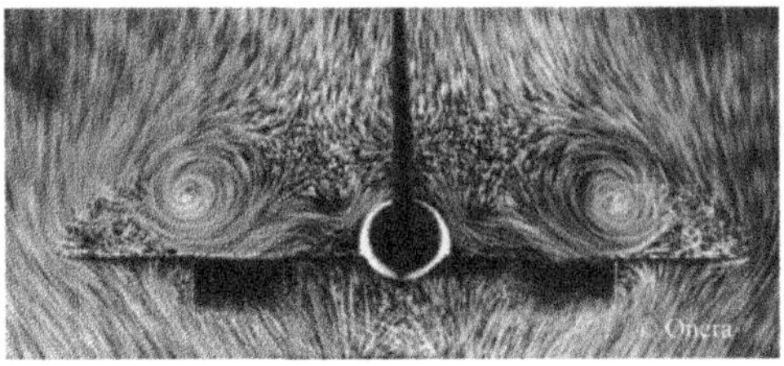

Figure 4.10. *Water tunnel visualization of the vortices over Concorde's wing's upper surface. Section by a plane perpendicular to the wing*

Because of its rapid swirl, the primary vortex induces an important depression over the wing's upper surface that is at the origin of the *vortex lift* giving a substantial increase in lift at high incidence (see Figure 4.11). Most supersonic aircraft utilize vortex lift for landing.

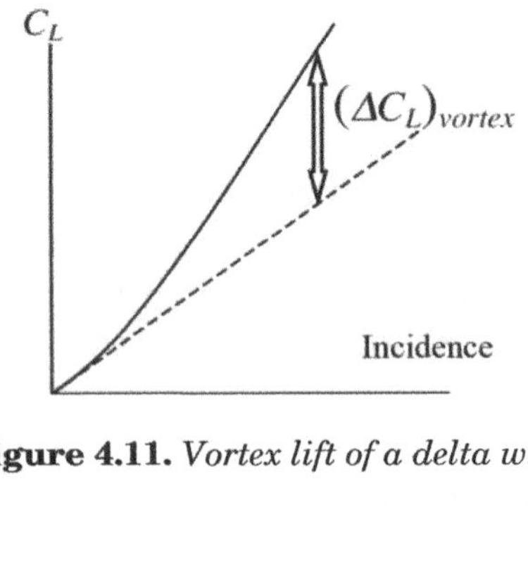

Figure 4.11. *Vortex lift of a delta wing*

4.2. Vortical flow over a slender body

A slender body is a long elongated body with a pointed nose as for the fuselage of a missile or of a combat aircraft. We will admit that the body includes a long cylindrical part (with a circular section) continuously connected to the sharp nose. The body length is such that phenomena in the base region do not affect the flow examined here. The body is set at an angle of incidence α in a uniform flow whose velocity $\vec{V}_\infty$ is contained in the symmetry plane.

The situation at the nose is similar to the situation at the delta wing apex if we consider that the nose presents some bluntness (a perfectly sharp nose cannot exist). Consequently, according to the results of the previous section, in the nose region, saddle points and nodes exist whose number depends on parameters such as the angle of incidence, the nose geometry and the Reynolds number. In most cases, two saddle points (S_1 and S_2) and three nodes (N_1, N_2 and N_3, N_1 being the main attachment node) are present. Two separation surfaces (S_1) and (S_2), whose wrapping forms two vortices, are associated with the saddle points.

The corresponding skin friction line pattern is shown in Figure 4.12. In the nose region, the two separation lines take a direction nearly parallel to that of a generator of the fuselage. On the windward side symmetry plane, there is an attachment line from which the skin friction lines diverge to turn around the fuselage and pass on the lee-side. This motion is caused by the pressure difference between the windward side (at higher pressure) and the lee-side (at lower pressure). After turning, the skin friction lines take a direction nearly parallel to the separation line (S_1). A converging or diverging motion, relative to a separation line, is due to the action of external forces (pressure and shear) and is a consequence of the fact that skin friction lines cannot cross (except at a critical point). Hence, any convergence must be along a separation line. In the present case, the convergence of the skin friction lines starts slightly before the end of the front nose. Then the motion accelerates, the skin friction lines coming from the windward side being abruptly deflected when approaching (S_1). Because of this motion, some authors state that separation may occur without existence of a saddle point in the surface flow pattern. It is then difficult to understand how a vortex can form without originating at a focus located on a material surface (Helmholtz's theorem).

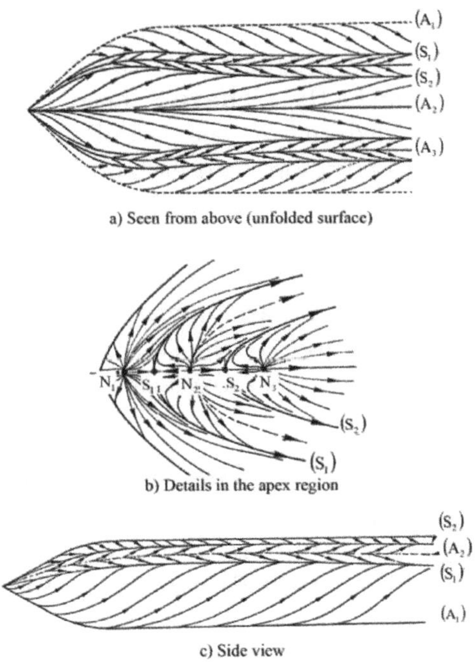

Figure 4.12. *Vortices over a slender body. Surface flow pattern*

The symmetry plane of the lee-side contains the attachment line (A_2) from which the skin friction lines diverge to tend toward the second separation line (S_2) according to a very progressive motion. The attachment line (A_3), which must exist between (S_1) and (S_2) comes from the node N_3 and demarcates the skin friction lines flowing either to (S_1) or to (S_2). The more or less rapid converging motion toward the separation line is dictated by the intensity of the forces acting in the fluid, in particular the pressure. Going from the windward side to the lee-side, the rise is pressure when passing round the cylinder induces a rapid convergence of the skin friction lines. On the other hand, the pressure being almost constant on the lee-side, the skin friction lines deflection is very progressive.

As shown in Figure 4.13, intersection of the two vortices by a plane normal to the fuselage axis shows two foci F_1 and F_2. By considering a half fuselage, there are five half-saddle points on the wall: S_1 on the windward side, S_2 on the lee-side, S_3 the trace of the separation line (S_1), S_4 the trace of the attachment line (A_3) and S_5 the trace of the separation line (S_2). In the field, there is the saddle S_6 point through which the separation line runs, splitting the pseudo-streamlines that spiral into the foci F_1 and F_2 from those that flow to infinity "above". When the angle of incidence is large ($\alpha \approx 20°$), the main vortex linked to the separation line (S_1) is well detached from the body, the second vortex being smaller and less intense.

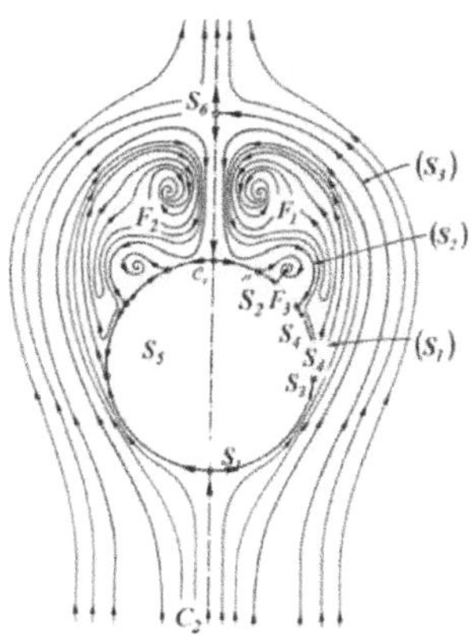

Figure 4.13. *Vortices over a slender body. Field projected in a plane normal to the body axis*

As for a delta wing, systems with more than two vortices develop according to the flow conditions. Figure 4.14 shows a system with three vortices (in a half plane) containing four attachment lines. There exists configurations with still more vortices. Limit circles are also observed (see Figure 4.15), the inner part of the vortex expanding (focus F_1 is said to be instable). The vortex fields of a slender body and a delta wing are topologically equivalent (same critical points and associated separation lines), the two fields differing by the arrangement, intensity and shape of the vortices and also by the surface flow properties.

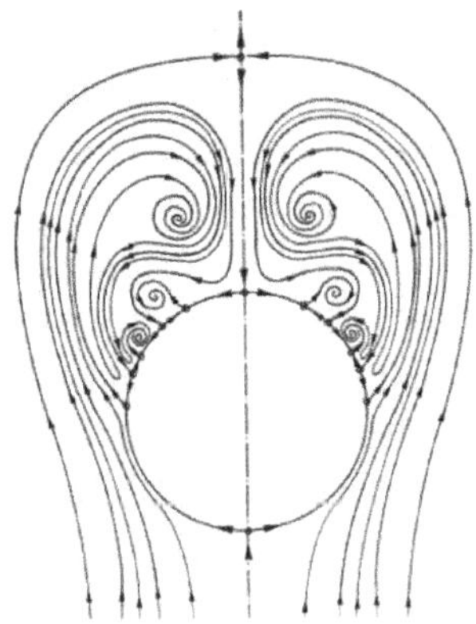

Figure 4.14. *Vortices over a slender body. Projected field. Three-vortex system*

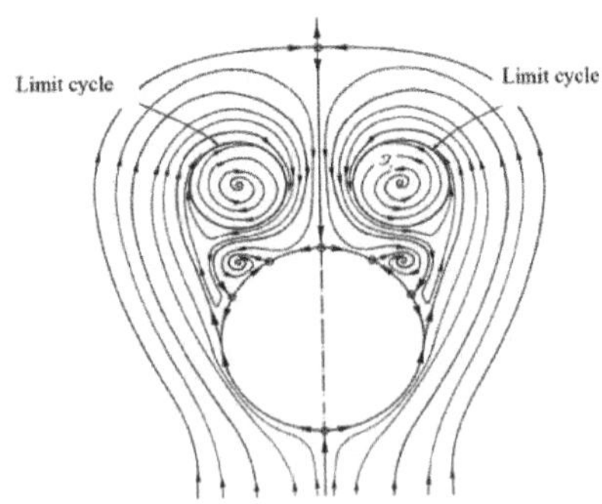

Figure 4.15. *Vortices over a slender body. Projected field. Existence of a limit cycle*

Figure 4.16 shows the vortices above a slender body set at an incidence in a Mach 2 supersonic flow. The vortices are visualized by injecting fine droplets of water that are illuminated by laser sheets. The vortex cores are visible as “black holes” due to the centrifugal effect that ejects the droplets from the vortex centers.

Figure 4.16. *Vortices over a slender body at an incidence in a Mach 2 supersonic flow*

At high incidence, because of the interaction between the vortices, the system made up of the two main (or primary) vortices does not remain symmetric (the secondary vortices play a negligible role in the phenomenon). The structure adopts the asymmetrical configuration shown in Figure 4.17 where the left-hand side vortex positions itself above the other vortex. The topology of the projected field is the same as in the symmetrical case, with the same number of critical points and separation lines. The phenomenon is illustrated in Figure 4.18 by water tunnel visualizations. Such an asymmetric pattern has important consequences for the guidance of a missile. The pressure being lower under the vortex closest to the surface, there results a side force making the control of the missile trajectory delicate. This phenomenon is also present at the nose of the fuselage of a combat aircraft maneuvering at an high incidence.

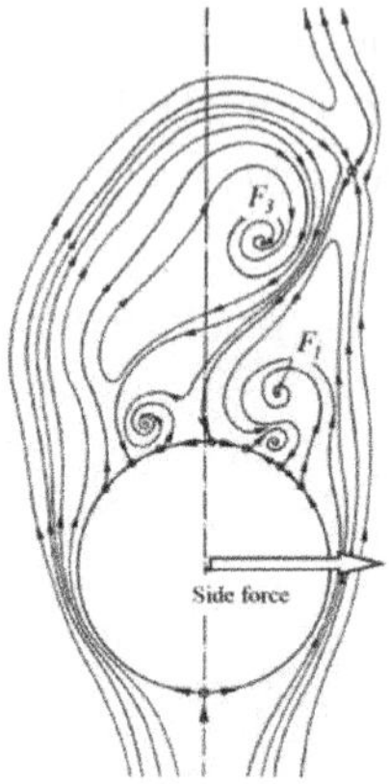

Figure 4.17. *Unsymmetrical vortex system over a slender body*

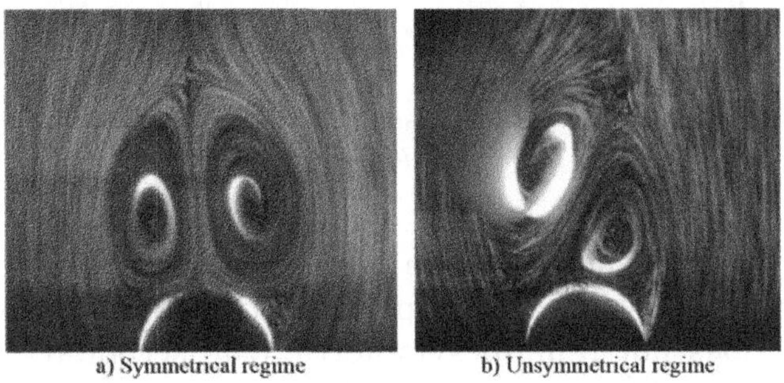

Figure 4.18. *Water tunnel visualization of the vortices over a slender body*

4.3. Vortex wake of a classical wing

4.3.1. *Topological description*

Let us consider a classical wing, with a slightly swept leading edge, set at an incidence without yaw in a uniform flow (the fuselage is ignored). The surface flow pattern corresponding to the simplest situation is represented in Figure 4.19. As shown in Figure 4.19(a), the attachment node N_1 at the wing's apex is the origin of all the skin friction lines enveloping the wing. The attachment line (A_1), running along the leading edge, splits these lines into a family flowing on the wing's lower surface and a family flowing on the wing's upper surface. On the upper surface, the skin friction lines, which are split into two families by the separation line (A_2), flow either to the left or to the right wing tip (see Figure 4.19(b)). As the pressure is higher on the lower surface, the skin friction lines of the lower surface are deflected toward the upper surface. There is a saddle point S_1 on the wing's trailing edge in the symmetry plane through which the separation line (S_1) running along the trailing edge flows (see Figure 4.19(c)). Near the wing's extremities, the separation line (S_1) spiral into the foci F_1 and F_2 into which they disappear. As shown in Figure 4.20, the separation line (S_1) sustains the separation surface (Σ_1), which rolls up to constitute the two tip vortices, whose traces on the wing are F_1 and F_2. The streamlines of (Σ_1) emanate from the separation node N_1 coincident with the saddle point S_1. The separation surface (Σ_1) constitutes the "skeleton" of the wake vortex of the wing.

As the wing is an isolated and simply connected body, we can apply equation [3.1] with:

$$\sum nodes + foci - \sum saddle \quad points = 1 + 2 - 1 = 2$$

which is well verified by this simple example.

Structures that are more complex exist, keeping in mind that the existence of the tip vortices implies the presence of two foci in the surface flow pattern. Except at a very low Reynolds number, the foci F_1 and F_2 are very small and close to the trailing edge, which makes their observation difficult. To satisfy Helmholtz's theorem, the tornado-like vortices emanating from the wing tips must either terminate on a material surface or close on them. In reality, the vortex structure is organized as represented in Figure 4.21, the tip vortices being connected to the starting vortex forming when the wing accelerates from rest.

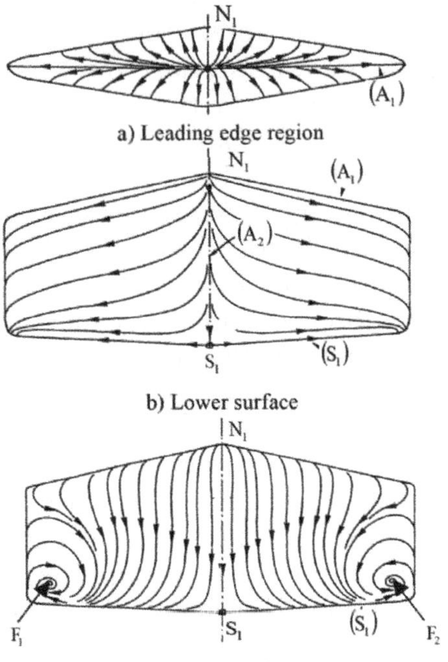

Figure 4.19. *Wake vortex of a classical wing. Surface flow pattern*

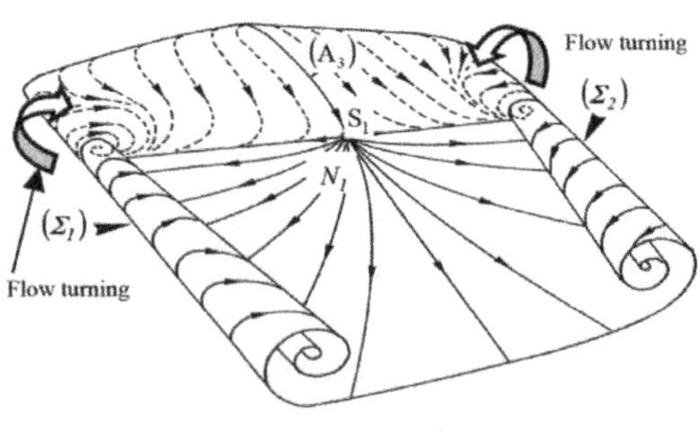

Figure 4.20. *Wake vortex of a classical wing. Separation surface and vortices*

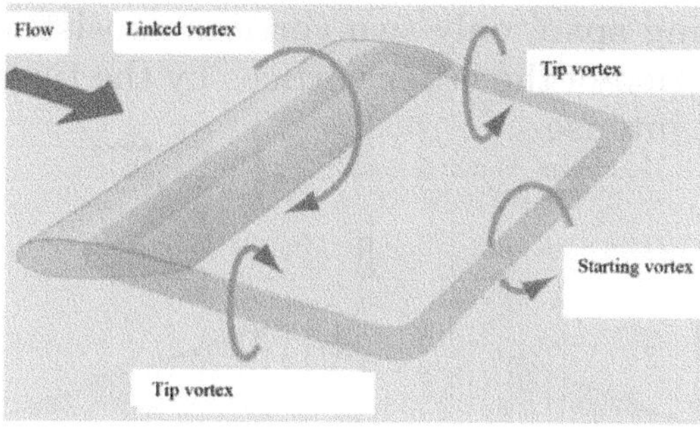

Figure 4.21. *Tip vortices and starting vortex of a wing*

The turning past the wing extremities by the lower surface flow and the resulting organization of the flow field is at the origin of the *induced drag*, which represents about 40% of the total drag for a modern transport aircraft. Figure 4.22 shows the surface flow pattern of a wing equipped with control surfaces (the reader can interpret the picture as an exercise). The counting of the critical points is as follows:

– number of nodes: 1 (attachment node in the front part);

– number of foci: 10;

– number of saddle points: 9.

Hence:

$$\sum nodes + foci - \sum saddle\ points = 1 + 10 - 9 = 2$$

The surface pattern being topologically consistent, the flow represented is possible. The separation surfaces and vortices escaping from the extremities of the wing and control surfaces are represented in Figure 4.23 and the flow is projected in a downstream plane in Figure 4.24. Downstream of the wing, the vortices interact between them and coalesce so that at a distance of approximately three times the wing span, only two vortices subsist as the result of the absorption of the other vortices by the tip vortices that are the most intense.

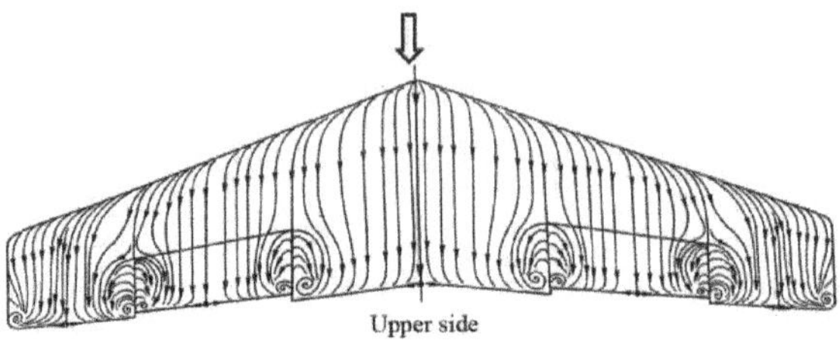

Figure 4.22. *Wake vortex of a wing with flaps. Surface flow pattern*

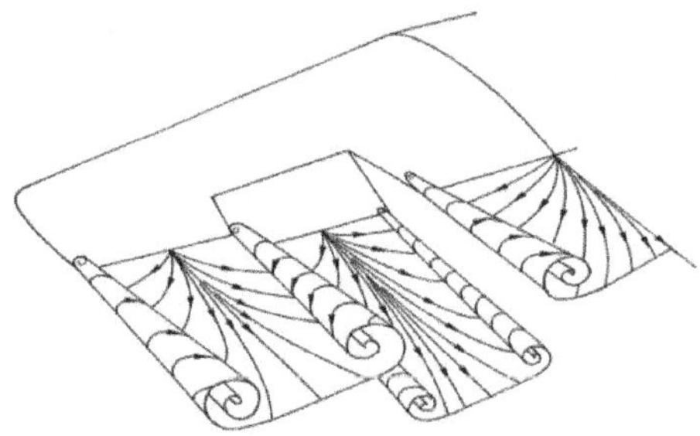

Figure 4.23. *Separation surfaces and vortices downstream of a wing with flaps*

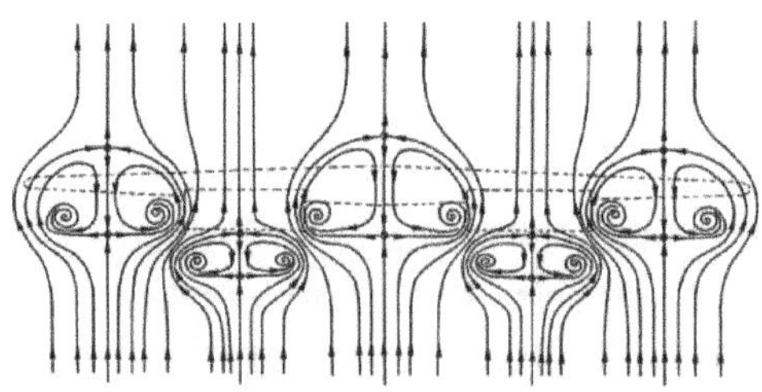

Figure 4.24. *Wake vortex of a wing with flaps. Field in a transverse downstream plane*

This phenomenon is illustrated in Figure 4.25 showing stagnation pressure distributions measured in the wake of an Airbus A300 model. Such distributions visualize the vortical structures since their core is a "hole" of the stagnation pressure (see section 2.5). After a distance of about three wingspans, only the tip vortices remain that will persist over a very long distance before dissipation under the action of viscosity. For large transport aircraft, the swirling motion induced by the tip vortices is so strong that it can destabilize a plane following at too a short distance. The definition of the safety distance between two successive takeoffs and landings takes this fact into consideration.

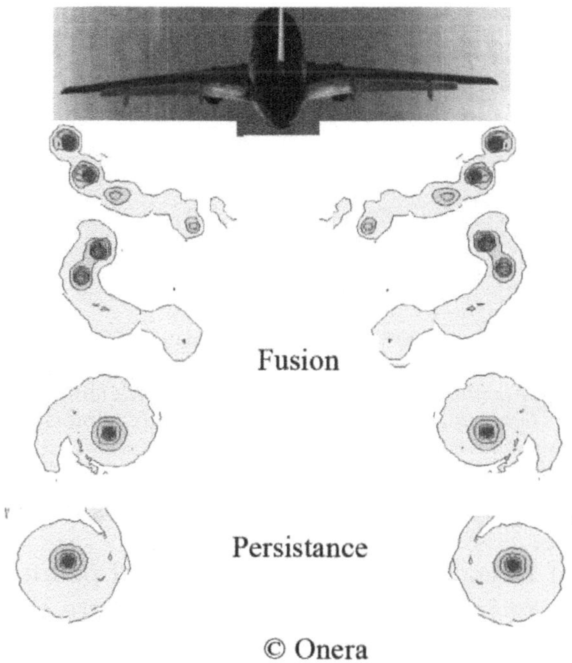

Figure 4.25. *Coalescence of vortices downstream of an Airbus A300 model in free flight with flaps*

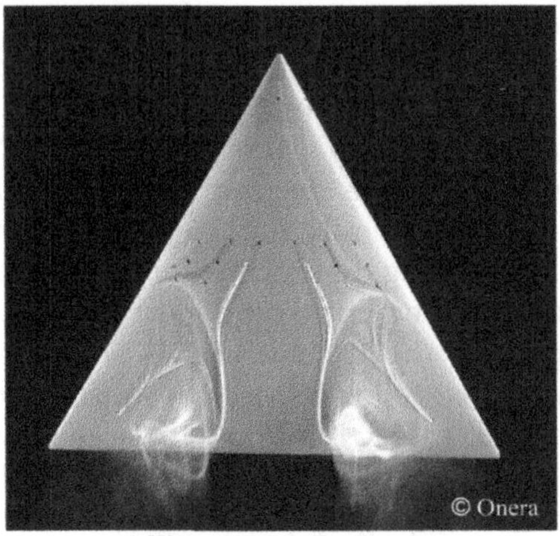

a) Water tunnel visualization

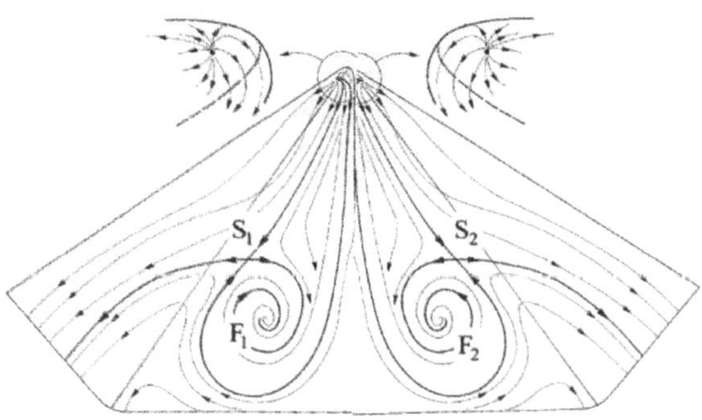

b) Topological interpretation

Figure 4.26. *Formation of the vortices over a delta wing with rounded leading edges at a very low Reynolds number*

4.3.2. *A scenario for the origin of vortices on a wing*

At usual Reynolds numbers, the critical points at a delta wing apex are indiscernible because these are confined in a very small region. A water tunnel experiment at a very low Reynolds number with a delta wing having a thick leading edge has allowed simulating an enlargement of the apex region. Figure 4.26(a) shows the existence of two foci and two saddle points on the wing's upper surface, the drawing in

Figure 4.26(b) (where the wing's lower part is unfolded) giving a topological interpretation of the surface flow pattern. The two distinct foci F_1 and F_2 on the wing's upper surface are the traces of tornado-like vortices escaping in the flow field. If the wing's thickness is progressively reduced (or the Reynolds number increased), we can imagine an evolution of the topology to the organization shown in Figure 4.1. The pattern constituted by the foci F_1 and F_2 and the saddle points S_1 and S_2 moves toward the apex so that the tornado-like vortices take on a common origin to reorganize themselves into a horseshoe vortex.

Considering Figure 4.26, another evolution of the topology can be conceived when the wing shape changes progressively from a delta to a rectangular form. With the geometrical modification, each focus moves to the wing's extremity where it is at the origin of the tip vortices of the classical wing (see water tunnel visualization and sketch in Figure 4.27).

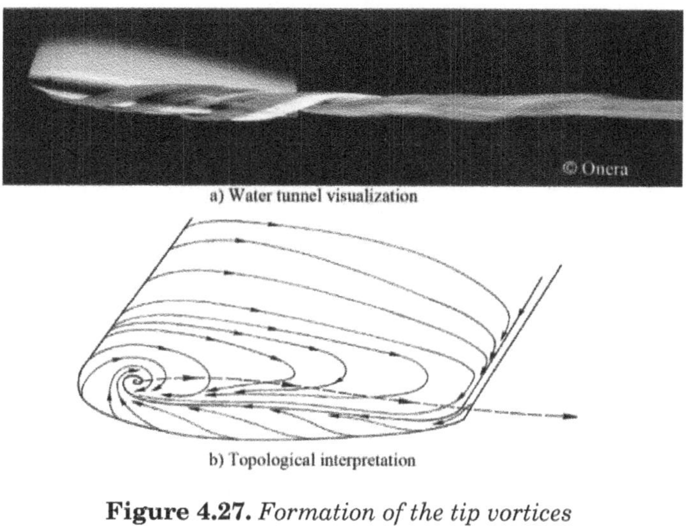

Figure 4.27. *Formation of the tip vortices*

5

Separation Induced by an Obstacle or a Blunt Body

The system of horseshoe vortices induced by separation ahead of a blunt obstacle is depicted. Then, the case of an obstacle of finite height (or protuberance) is considered. The topologies of the flow downstream of a missile-type body at incidence with a base and of the flow past a sedan type automobile are investigated, including the description of the emitted vortex wake.

5.1. Separation in front of an obstacle

Frequently, the flow encounters a blunt obstacle more or less inclined with respect to the upstream velocity vector. This is the case of a vertical tail, the pylon of a nacelle, a submarine sail, a bridge pier or the radiator grille of an automobile. To simplify, we consider a cylindrical obstacle, with a rounded front part, mounted without incidence, normally to a flat wall, and the upstream flow being uniform. The obstacle is sufficiently high and long so that the phenomena occurring at its trailing edge and upper part play a negligible role.

The simplest separated configuration induced by such a geometry leads to the surface flow pattern shown in Figure 5.1 (also see the visualization in Figure 1.7). In front of the obstacle, there is the separation line (S_1) associated with the saddle point S_1, with the upstream skin friction lines turning abruptly as they approach (S_1). According to their position relative to the second separation line (S_2), they turn either to the left or to the right of the obstacle. The surface flow downstream of (S_1) is fed by the node N_1 from, which emanates the skin friction lines flowing either to (S_1) or in the downstream direction, the two families being separated by the attachment line (A_1). The turning motion of the skin friction lines downstream of (S_1) is much more progressive than upstream.

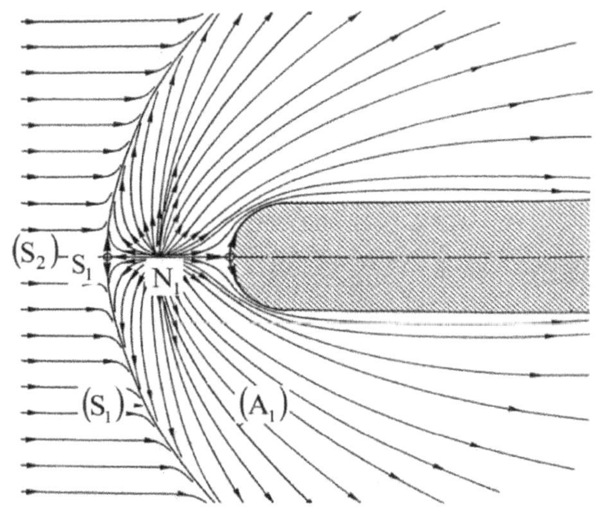

Figure 5.1. *Separation induced by an obstacle. Surface flow pattern in front of the body*

The flow in a vertical symmetry plane is represented in Figure 5.2. The separation line (S_1), coming from the half-saddle point S_1 coincident with the saddle point S_1 of the surface flow pattern, winds into the focus F_1 that is the trace in the symmetry plane of the vortex resulting from the separation induced by the obstacle. The streamlines going into F_1 are comprised between (S_1) and the separation line

(S_2) terminating at the half-saddle S_2 coincident with the node N_1. Downstream, different organizations are observed depending on the flow conditions (the Reynolds number in particular). In the case shown in Figure 5.2(a), the streamlines above (S_2) flow along the obstacle, the considered region containing only three critical points (a focus and two half-saddle points). In the case of Figure 5.2(b), a second separation occurs in front of the obstacles to which corresponds the separation line (S_2) of the surface pattern, and in the symmetry plane the separation line (S_3) winding around the focus F_2 trace of a second vortex. A fourth separation line (S_4) ends at the half-saddle point S_4 on the obstacle. The flow in the symmetry plane contains six critical points (two foci and four half-saddle points). For the configuration shown in Figure 5.2(d), the second separation takes place on the obstacle.

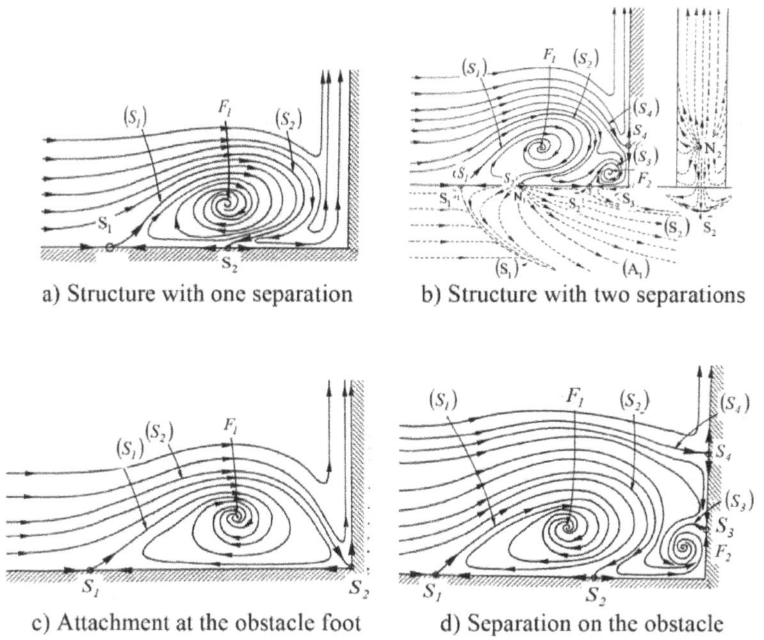

Figure 5.2. *Separation induced by an obstacle. Flow in the symmetry plane: variants*

Figure 5.3 shows the separation surface (Σ_1) sustained by the separation line (S_1). Its streamlines are issued from the node N_1 coincident with the saddle point S_1. Here, the separation surface (Σ_1) wraps to constitute a horseshoe vortex surrounding the obstacle. At low Reynolds numbers, especially if the boundary layer is laminar, several separations take place.

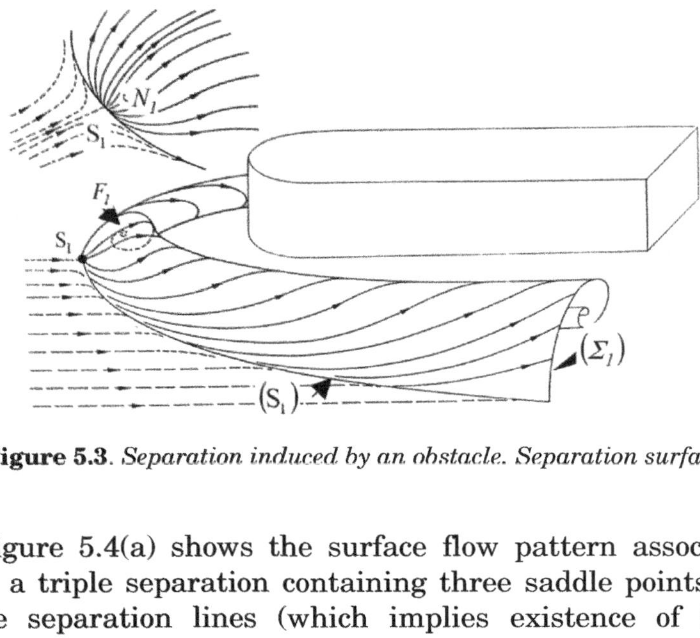

Figure 5.3. *Separation induced by an obstacle. Separation surface*

Figure 5.4(a) shows the surface flow pattern associated with a triple separation containing three saddle points and three separation lines (which implies existence of three attachment lines). The flow in the symmetry plane represented in Figure 5.4(b) contains four foci and four vortical structures (along with a fourth separation on the obstacle). Figure 5.5 shows the separation surfaces associated with three main separations and Figure 5.6 shows a water tunnel visualization of the separation induced by a cylindrical obstacle. The tracer lines of dye emitted upstream, very close to the surface of the plate, reveal a separation with one saddle point and one separation line acting as a barrier for wall flow. A tracer line emitted downstream is captured by the horseshoe vortex formed by the separation surface.

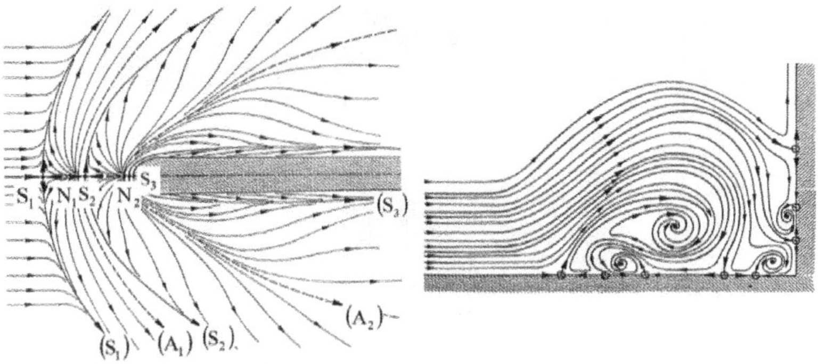

a) Skin friction line pattern b) Flow in the symmetry plane

Figure 5.4. *a) Separation induced by an obstacle with three separation lines. b) Surface flow pattern*

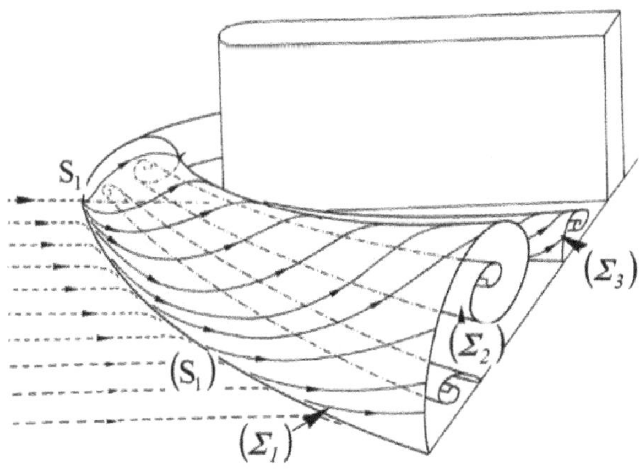

Figure 5.5. *Separation induced by an obstacle with three separation lines. Separation surfaces*

Organization for a two-vortex system projected in a plane normal to the flat surface is represented in Figure 5.7. Considering one-half of the figure, we observe two foci, traces of horseshoe vortices, two attachment lines (one of them being in the symmetry plane), two separation lines of the separation type and four half-saddle points at the wall (one being in the symmetry plane).

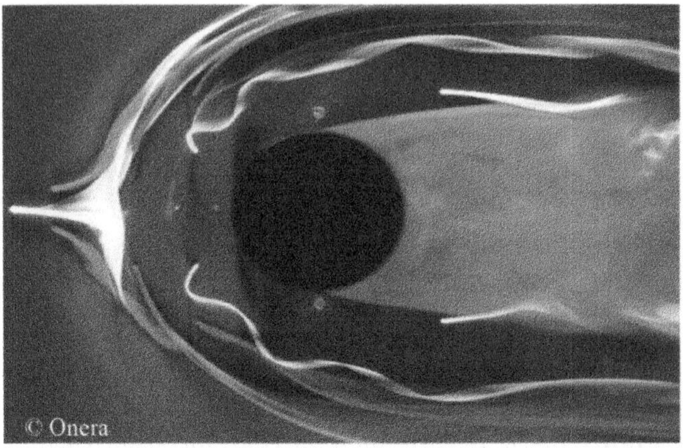

Figure 5.6. *Water tunnel visualization of the separation induced by a cylindrical obstacle*

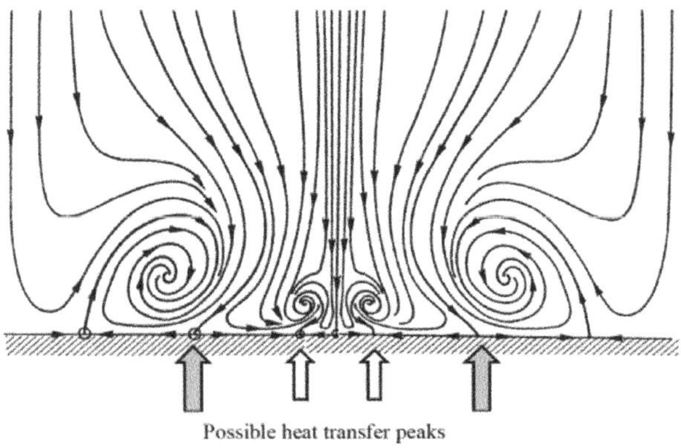

Figure 5.7. *Separation induced by an obstacle with two separation lines. Field projected in a downstream plane*

The pseudo-streamlines oriented from above disappear in the foci. The half-saddles (indicated by arrows) are attachment or stagnation points for the flow in the vertical plane. In high-speed flows, these points are regions of heat transfer peak due to the high stagnation enthalpy level of the outer flow entrained to the wall by the vortex motion. When the flight Mach number is above five, the local heat transfer levels can damage the fuselage of missiles or space launchers supporting different kinds of protuberances (the housing cover of antennas, pipes, etc.).

5.2. Flow induced by an obstacle of finite height or protuberance

Let us consider a cylindrical obstacle, with a circular cross-section, normal to a flat plate placed in a uniform upstream flow. If the height of the obstacle is comparable to the diameter of the cylinder cross-section, it constitutes a protuberance. Figure 5.8 shows a viscous film visualization of the surface flow pattern induced by a cylinder in a Mach 2.5 supersonic flow. The two identified traces are due to the shock waves induced by the obstacle: one trace corresponds to the detached shock in front of the obstacle and the other trace corresponds to the downstream leg of the lambda structure resulting from the separation. These shocks made visible by their dioptre effect do not extend to the surface and do not directly interfere with the surface flow.

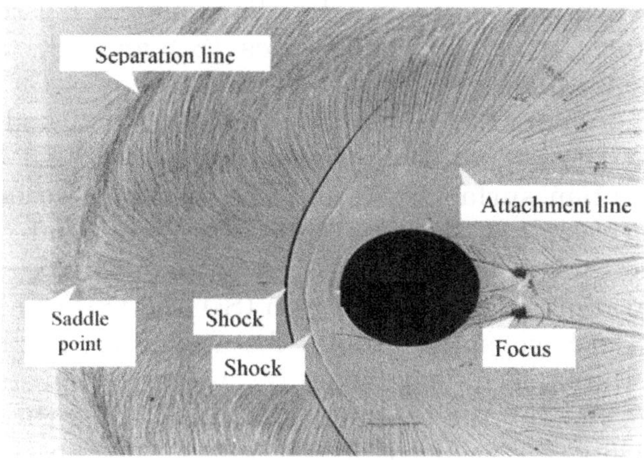

Figure 5.8. *Visualization of the surface pattern of the flow induced by a cylindrical obstacle in a supersonic stream [SED 58]*

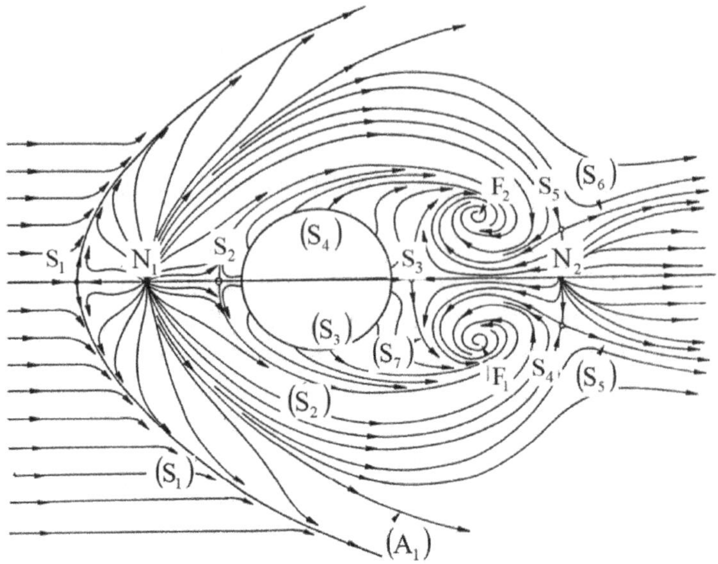

Figure 5.9. *Flow induced by a protuberance. Surface flow pattern on the plate bearing the obstacle For a color version of this figure see www.iste.co.uk/delery/3Dflow.zip*

Figure 5.9 shows the topological interpretation of the surface flow pattern on the flat plate. Upstream of the obstacle, two separation lines (S_1) and (S_2) and one attachment line (A_1) form. On the line of symmetry, the saddle points S_1 and S_2 and the attachment node N_1 are located. The separation lines (S_1) and (S_2) sustain the separation surfaces (Σ_1) and (Σ_2) forming two horseshoe vortices (we admit that conditions are such that only two separation lines are present). Downstream of the cylinder, the separation lines (S_2), and also (S_3) and (S_4) coming from the cylinder base, spiral into the foci F_1 and F_2. The backflow fed by the attachment node N_2 is separated from the upstream flow by the separation line (S_7) going through the saddle point S_3. Downstream of the foci, on each side of the symmetry axis, there are two saddle points S_4 and S_5 through which the separation lines (S_5) and (S_6) pass. One branch of each of these lines spirals into the focus F_1 or F_2, and the other branch extends downstream. These lines separate the skin friction lines disappearing into the foci from those flowing downstream.

As shown in Figure 5.10, the surface flow pattern on the obstacle contains an attachment node N_3 on the leading edge from which the skin friction lines wrapping the obstacle start. These lines converge to the separation line (S_8) (going through the saddle point S_6, which is the continuation of the obstacle of the separation line (S_3)), the flow downstream of (S_8) being fed by the node N_4. The second separation line (S_9) going through S_6 divides the skin friction lines into a set flowing toward the obstacle base and a set flowing to the obstacle top (the same organization exists on the other side of the cylinder). The skin friction lines above (S_8) are evacuated in the separation node N_5. The separation line (S_8) follows the top edge of the obstacle to end in the node N_5. In reality, a more complex flow must exist on the obstacle top, but due to the lack of information, we consider, here, the simplest solution. The numbering of the critical points is as follows:

– five nodes, plus one at upstream infinity and one at downstream infinity;

– two foci;

– seven saddle points.

Hence:

$$\sum nodes + foci - \sum saddle\ points = 7 + 2 - 7 = 2$$

The pattern represented is topologically admissible, which does not mean that the represented flow is physically possible.

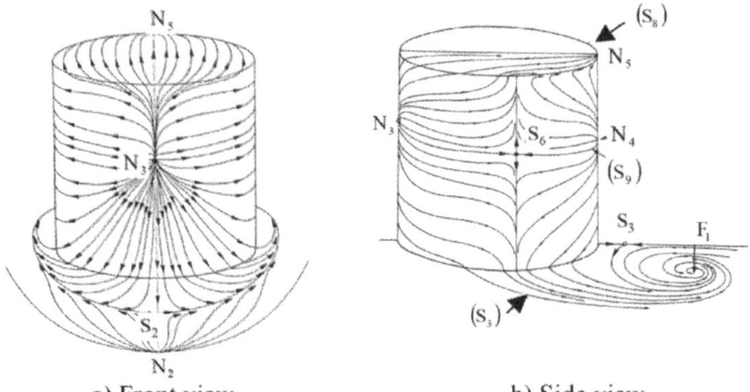

Figure 5.10. *a) Flow induced by a protuberance. b) Surface flow pattern on the obstacle*

The separation line (S_3) and its symmetric (S_4) sustain the separation surface (Σ_3) forming two tornado-like vortices whose trace on the flat plate are the foci F_1 and F_2. The streamlines constituting the stream surface (Σ_3) originate at the node N_3 coincident with the saddle point S_6. As shown in Figure 5.11, these two vortices bend in the downstream direction under the main stream action. These vortices are intense rotating structures whose origin is the separation occurring slightly after the obstacle maximum cross-section.

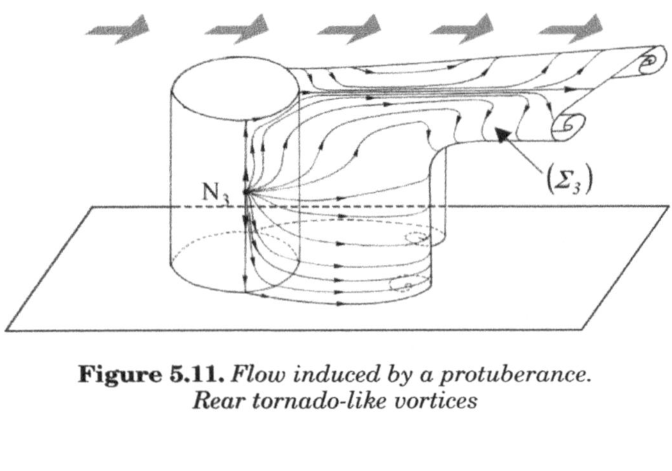

Figure 5.11. *Flow induced by a protuberance. Rear tornado-like vortices*

The water tunnel visualization in Figure 5.12 shows the flow induced by such an obstacle, details downstream of the cylinder being blurred by turbulence. However, we can distinguish the two tornado-like vortices that are deflected by the main flow. The analyzed case includes three separation surfaces giving rise to vortical structures as shown in Figure 5.13(a). A cut by a downstream vertical plane gives the organization as shown in Figure 5.13(b).

Figure 5.12. *Water tunnel visualization of the flow induced by a cylinder*

Tornado-like vortices can terminate on a free surface (see Figure 5.14) like downstream of a bridge pier (or of any partly submerged obstacle) whose vortices extend from the bottom of the river to its surface. Such vortices, responsible for the digging of the riverbed, can provoke the collapse of the pier. The aerodynamics of protuberances have a special importance because they are at the origin of drag, noise, local heat transfer peak (as seen above), etc.

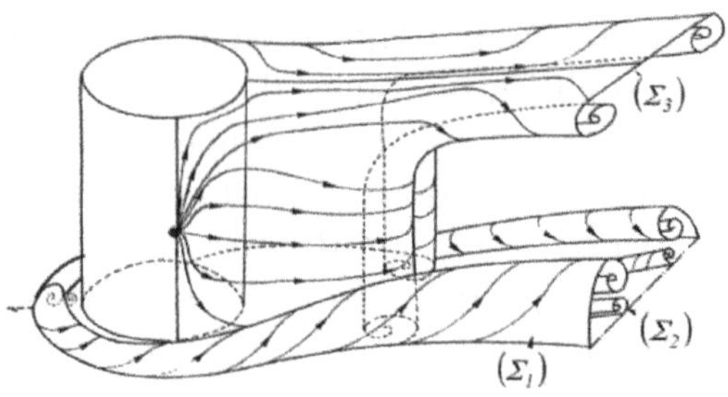

a) Separation surfaces

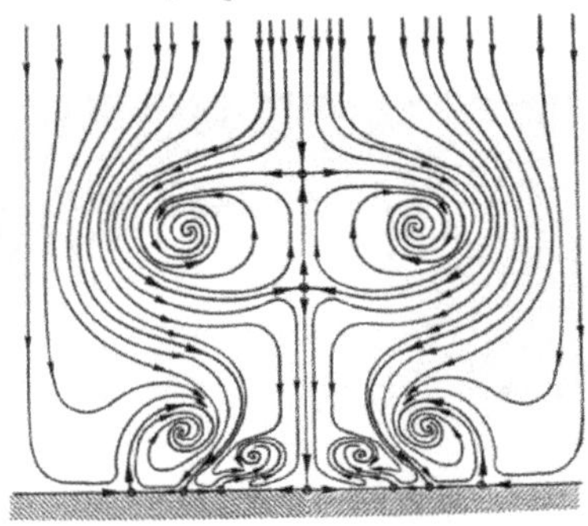

b) Field projected in a downstream plane

Figure 5.13. *Vortices induced by a protuberance immerged in a stream*

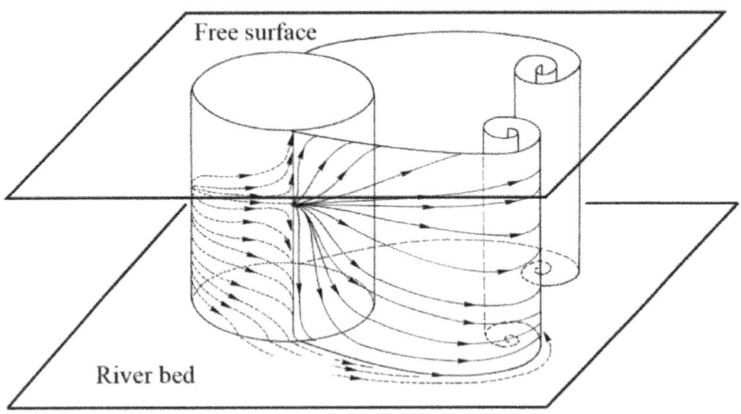

Figure 5.14. *Flow induced by a protuberance between the riverbed and the free surface. Rear tornado-like vortices*

5.3. Separation on a non-propelled afterbody

The simplest configuration is an axisymmetric body set at an incidence α in a uniform flow. Phenomena on the fore part of the body are ignored and we admit that the flow does not contain vortices produced by upstream separations. A base with a shoulder of a very small radius or a curvature terminates the afterbody.

On the surface flow pattern shown in Figure 5.15, the skin friction lines coming from upstream are divided into two sets on each side of the attachment line (A_1) located in the windward side plane of symmetry. The skin friction lines flow toward the separation line (S_1) located on the lee-side in the symmetry plane. When approaching the base, these lines turn rapidly to follow the base shoulder that contains the separation line (S_2) going through the saddle point S_1 located in the symmetry plane. All the body skin friction lines terminate at node N_1 located on the base shoulder on the lee-side (see Figure 5.16).

The lines originating from the attachment node N_2 located in the symmetry plane, closer to the windward side, flow toward the base and turn rapidly to follow the separation line (S_2) going through S_1; then they terminate in the separation node N_1. The critical points S_1 and N_1 belong to the surface flow pattern made up of the skin friction lines coming from upstream and those originating at N_2. Figure 5.17 shows a perspective view of the afterbody. If we admit that the upstream skin friction lines come from a unique node at the body nose, then we verify:

$$\sum nodes + foci - \sum saddle\ points = 3 - 1 = 2$$

and the flow is separated since its surface pattern contains more than two nodes (with one saddle point).

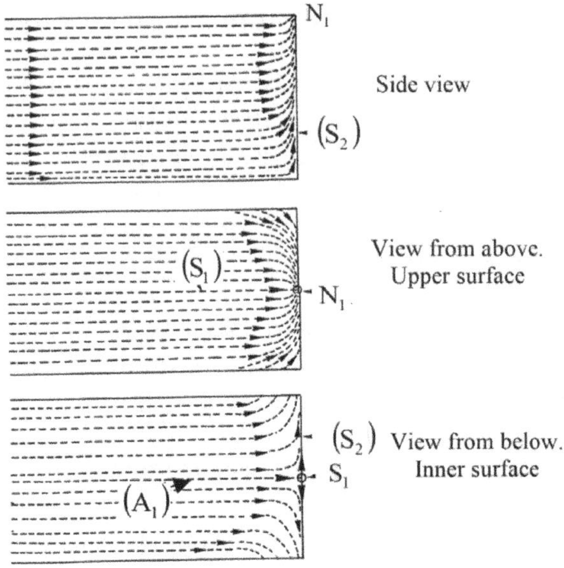

Figure 5.15. *Non-propelled afterbody. Surface flow pattern on the fuselage*

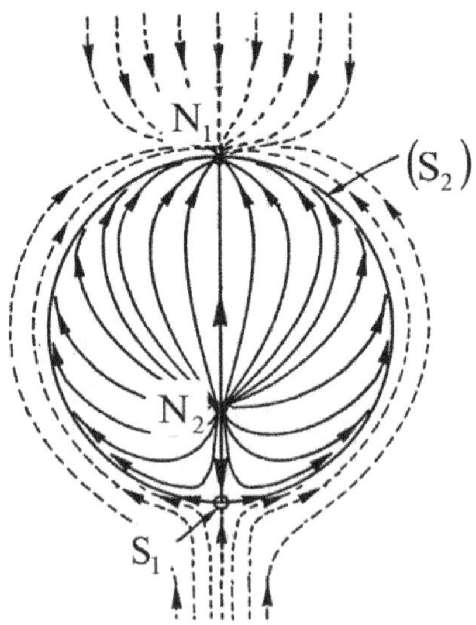

Figure 5.16. *Non-propelled afterbody. Surface flow pattern on the base. For a color version of this figure see www.iste.co.uk/delery/3Dflow.zip*

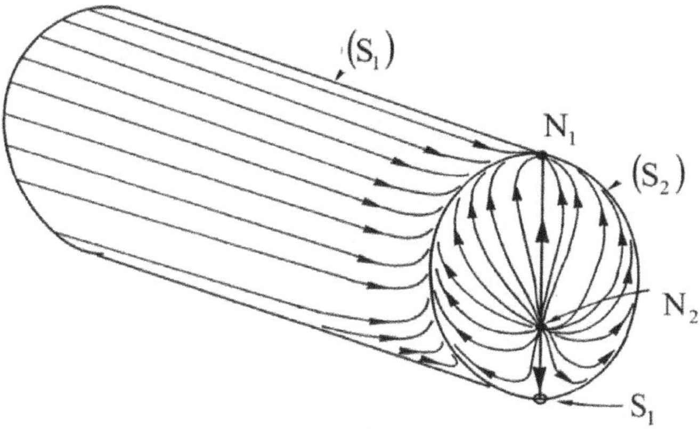

Figure 5.17. *Non-propelled afterbody. Surface flow pattern on the fuselage and the base*

As shown in Figure 5.18, the flow in the vertical symmetry plane contains the separation line (S_1), starting from A at the base shoulder on the lee-side, which spirals into the focus F_1. The point A is a half-saddle for this flow. The second separation line (S_2) ends at the half-saddle point S_2 located on the base and coincident with the node N_2. A third separation line (S_3) starts from the half-saddle S_3 located on the base shoulder lower part. The separation line (S_2) separates the streamlines spiraling into F_1 from those escaping downstream, the separation line (S_3) discriminating the streamlines coming from the windward and lee-sides. The streamlines located above (S_2) turn around it to flow toward the base; then they bend when approaching the saddle point S_2 and flow downstream along the separation line (S_3).

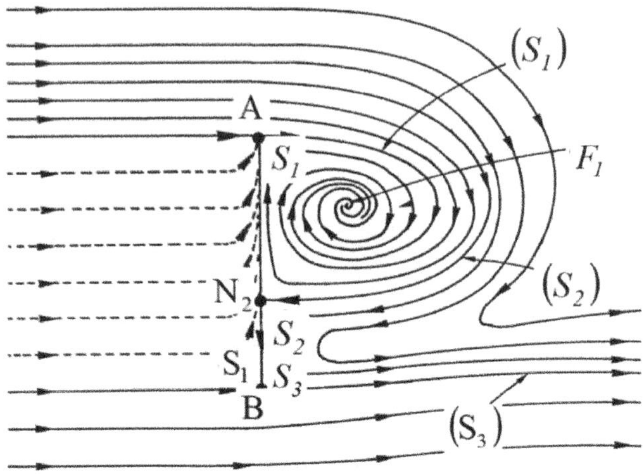

Figure 5.18. *Non-propelled afterbody. Flow in the vertical symmetry plane. For a color version of this figure see www.iste.co.uk/delery/3Dflow.zip*

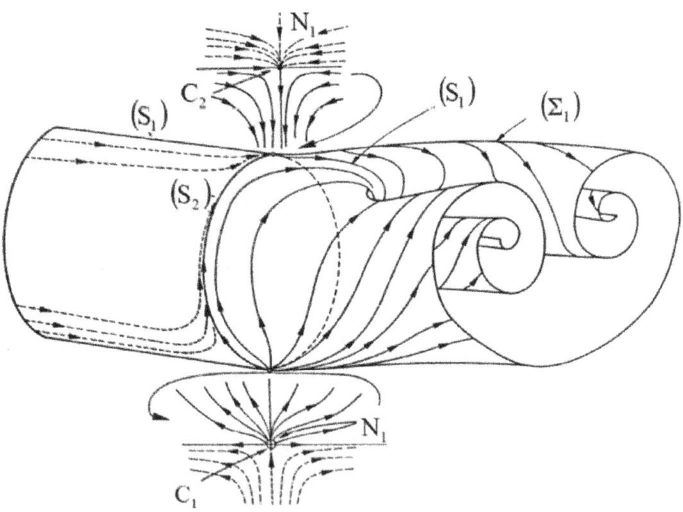

Figure 5.19. *Non-propelled afterbody. Separation surface*

As shown in Figure 5.19, the separation line (S_1) sustains the separation surface (Σ_1) whose streamlines originate at N_1

coinciding with the saddle point S_1. The surface (Σ_1) wraps into a horseshoe vortex whose legs take the downstream direction, the trace of this vortex in the symmetry plane being the focus F_1. The projection of the flow in a plane normal to the upstream velocity is shown in Figure 5.20. The trace of the separation surface (Σ_1) is the separation line (S_4) that spirals into F_2 and F_3. The flow pattern also contains two saddle points S_4 and S_5, two nodes at infinity "above" and "below", and the separation line (S_5) that discriminates the pseudo-streamlines trapped in the foci from those going from infinity "below" to infinity "above".

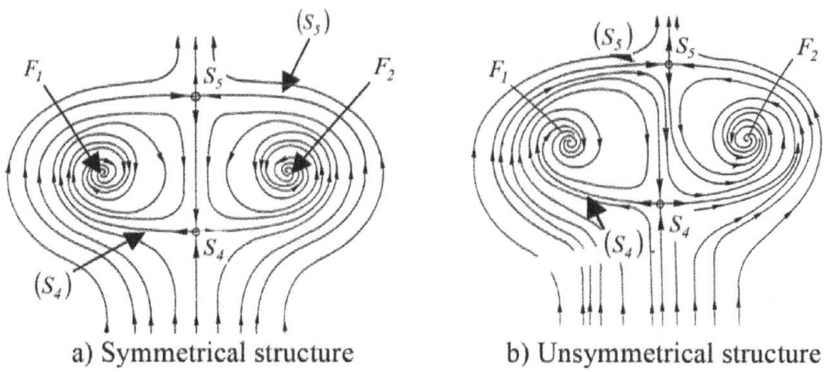

Figure 5.20. *Non-propelled afterbody. Field in a downstream projection plane*

As mentioned in section 3.4, a symmetrical pattern with two saddle points S_4 and S_5 located on the same separation line is unlikely. To the structure of Figure 5.20(a), we must substitute that of Figure 5.20(b) that is more plausible because of its less special character, the whole flow field being not much affected by this change.

If the angle of attack is high ($\alpha \geq 10°$), separation occurs on the front body part, as studied in section 3.2. If we assume (to simplify) that this front separation is unique (only one vortex system), the skin friction lines pattern is organized as shown in Figure 5.21 where the body surface has been unfolded. At the nose, there are the main attachment node N_3, the saddle point S_2 and the node N_4, the origin of the separation line (S_3) terminating at the node N_1. All the skin friction lines issued from N_3, N_4 and N_2 (on the base) terminate in the separation node N_1. The number of nodes exceeding by two the number of saddle points, the pattern is topologically consistent. As shown in Figure 5.22, the flow field contains two separation surfaces: (Σ_1) sustained by the separation line (S_2) along the base shoulder and (Σ_2) sustained by the separation line (S_3).

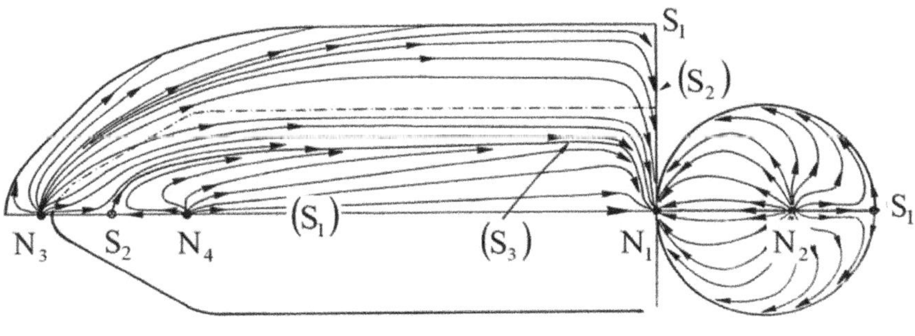

Figure 5.21. *Non-propelled afterbody at incidence. Surface flow pattern*

As shown in Figure 5.23, the field projected in a plane normal to the upstream velocity contains four foci that are the traces of the two horseshoe vortices of Figure 5.22.

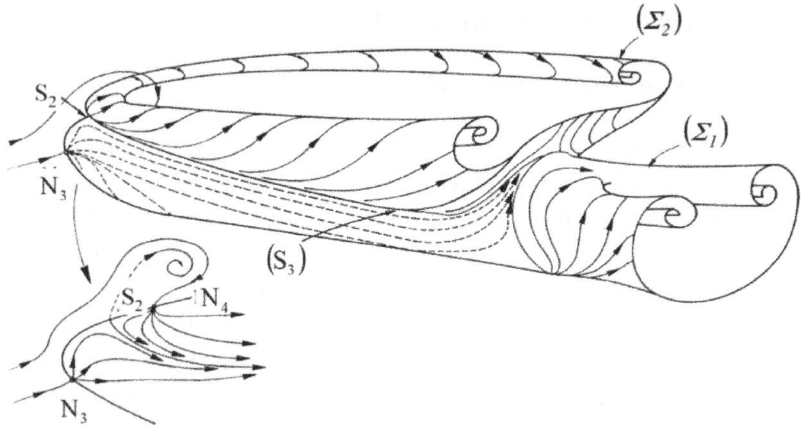

Figure 5.22. *Non-propelled afterbody at incidence. Separation surfaces*

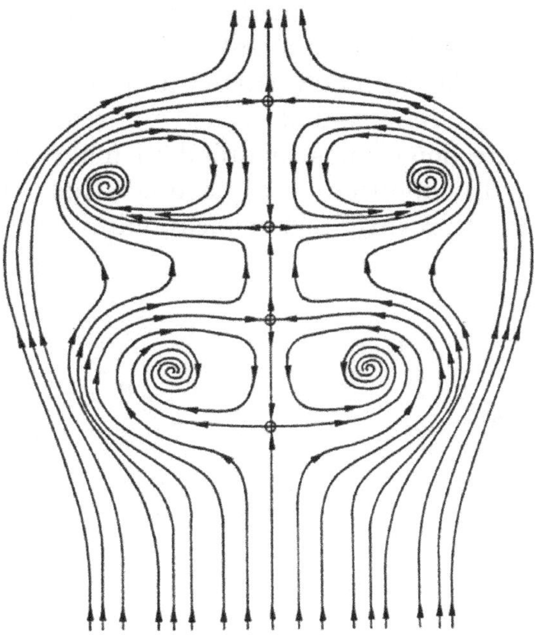

Figure 5.23. *Non-propelled afterbody at incidence. Field in a downstream projection plane*

5.4. The flow past an automobile

5.4.1. *The surface flow pattern*

We consider the flow past a three-body sedan type automobile including the bonnet in the front part, the passenger cell and the boot in the rear part. As shown in Figure 5.24, for the sake of simplicity, the vehicle geometry is made up of flat surfaces; there are no wheels or accessories (rear view mirrors, ferders, exhaust pipe, no holes, etc.), and the car's under-body is continuous. Consideration of a shape with rounded surfaces would have complicated the picture without influencing much the overall flow organization. However, the negligence of accessories can have more important consequences because they are the cause of vortex formation producing noticeable noise.

The following topological analysis is based on experiments including surface flow visualizations, flow field survey by Laser Doppler Anemometry (LDA) and field visualization. The structures represented exist in reality and their proportions are sometimes modified to make the drawings more clear. In addition, we assume that the flow is symmetrical with respect to a vertical plane (no yaw).

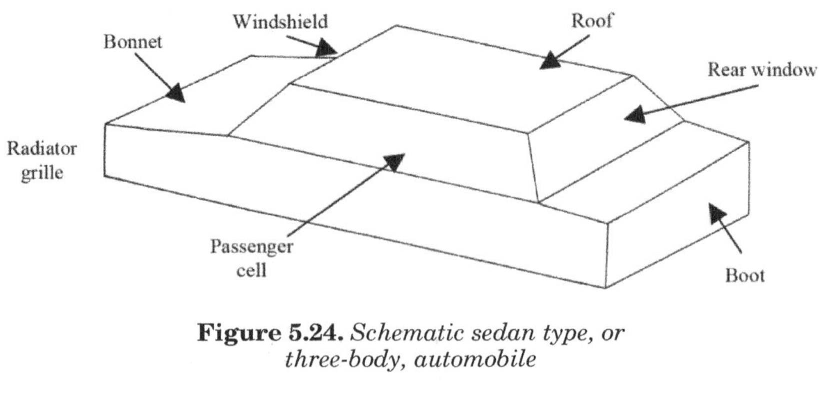

Figure 5.24. *Schematic sedan type, or three-body, automobile*

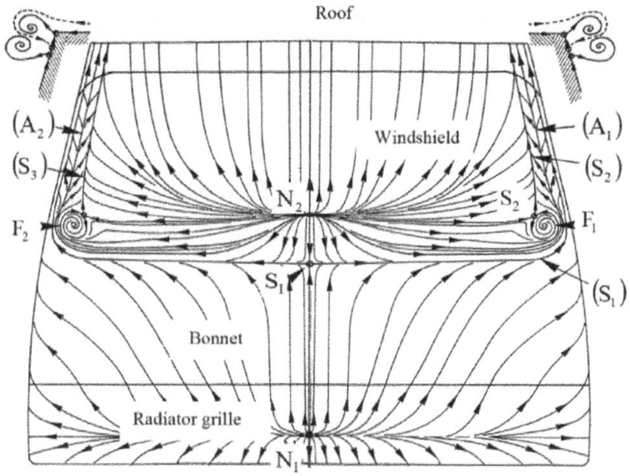

Figure 5.25. *Surface flow pattern on the automobile front part. For a color version of this figure see www.iste.co.uk/delery/3Dflow.zip*

Figure 5.25 shows a front view of the surface flow pattern containing the following features:

– a main attachment node N_1 is located on the radiator grille, the skin friction lines originating at N_1 flowing under the vehicle, on the bonnet and along the sides.

– on the bonnet, the separation line (S_1) going through the saddle point S_1 is the trace of the separation induced by the windshield (obstacle effect). This line flows laterally, then along the A-pillar and continues along the roof. The separation line (S_1) is the trace of the separation surface (Σ_1), the wrapping of which gives a horseshoe vortex propagating downstream after having followed the passenger cell.

– a second attachment node N_2, located on the windshield, is the origin of the skin friction lines flowing on the windshield and the roof.

– the turning around of the passenger cell induces on each side a separation with the saddle points S_2 and S_3 along with the separation lines (S_2) and (S_3). These lines spiral around the foci F_1 and F_3 that are the traces of tornado-like vortices following the A-pillars of the windshield and the edge of the passenger cell.

Between the separation lines (S_1) and (S_2), on the one hand, and (S_1) and (S_3), on the other hand, attachment lines (A_1) and (A_2) are present.

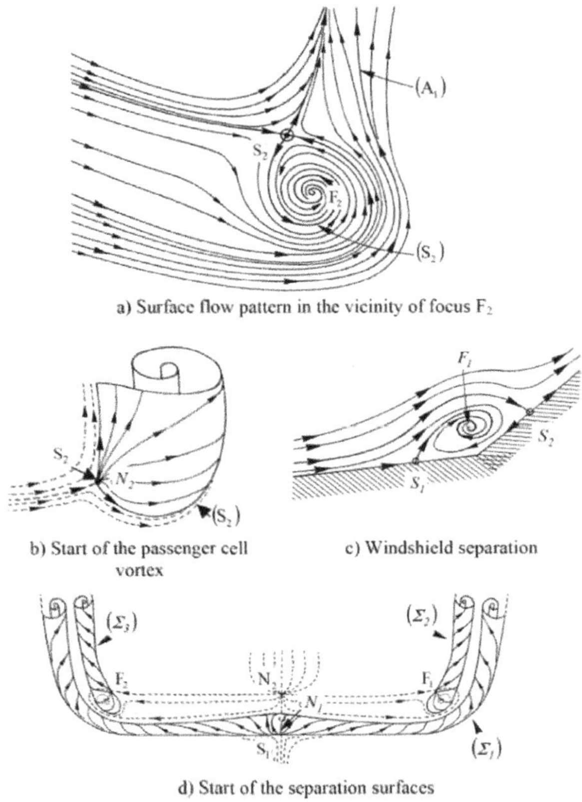

Figure 5.26. *Formation of the windshield and passenger cell vortices. For a color version of this figure see www.iste.co.uk/delery/3Dflow.zip*

Figure 5.26(a) shows the surface flow in the vicinity of the foci F_1 or F_3 and Figure 5.26(b) shows the start of the associated vortex. The separation surface forming this vortex is fed by node N_1, coincident with the saddle point S_2. The cut of the windshield vortex at the level of the saddle point S_1 (see Figure 5.26(c)) contains the focus F_1 and the two half-saddle points S_1 and S_2 where the separation lines stagnate. The two vortices formed by the wrapping of the separation surfaces (Σ_1) and (Σ_2) are drawn in Figure 5.26(d). At this step of the analysis, we observe two vortical structures: one associated with the windshield separation and the other with the separation that occurs when passing around the A-pillar.

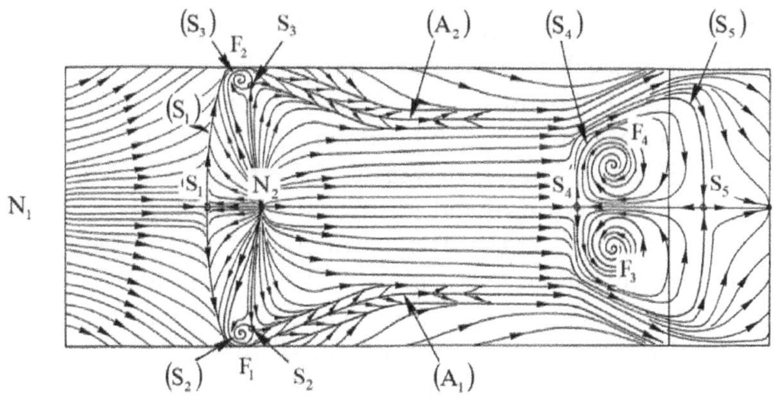

Figure 5.27. *Surface flow pattern seen from above. For a color version of this figure see www.iste.co.uk/delery/3Dflow.zip*

Figure 5.27 represents the surface pattern seen from above, which shows two foci F_3 and F_4 linked to the separation on the rear window and the separation line (S_4) spiralling around these foci. A second separation line (S_5) going through the saddle point S_5 demarcates the skin friction lines spiralling into F_3 and F_4 from those flowing on the car rear part. This classical pattern has been analyzed in section 2.4. Figure 5.28 shows the surface pattern on the automobile's side containing the attachment line (A_3), (another attachment line exists on the other side of the car).

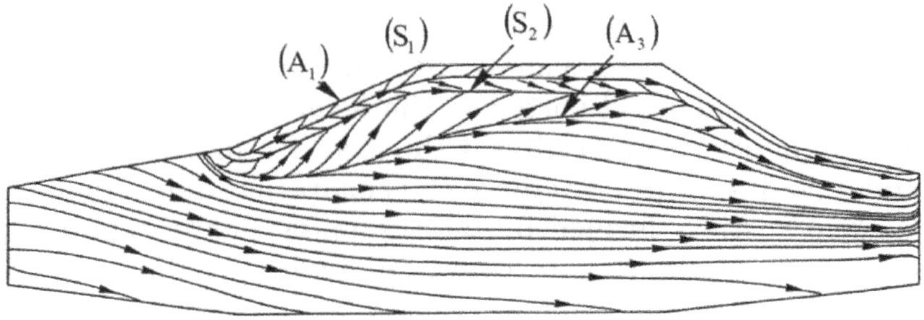

Figure 5.28. *Side view of the surface flow pattern*

Figure 5.29 shows the surface pattern on the vertical part of the boot that behaves like a base. The variant drawn in Figure 5.29(a) contains the nodes N_3 and N_4 and also the saddle point S_6 through which flows the separation line (S_6). The organization shown in Figure 5.29(b) shows three nodes and two saddle points located laterally. The separation lines (S_1), (S_2) and (S_3), which do not terminate in the foci, end in the separation nodes N_5 and N_6. The attachment lines (A_1) and (A_3), as well as the symmetrical lines (A_2) and (A_4), go through the saddle points S_7 and S_8. If the flow under the car does not contain critical points (which is possible in the absence of wheels or any holes or protuberances) and if the car has a surface of complexity zero (that is two windows are not open at the same time), then the following cases arise:

– Variant of Figure 5.29(a):

- saddle points: 8,
- foci: 4,
- nodes: 6.

Thus, relation [3.2] is verified because $(4+6)-8=2$.

– Variant of Figure 5.29(b):

- saddle points: 9,
- foci: 4,
- nodes: 7.

Hence, $(4+7)-9=2$.

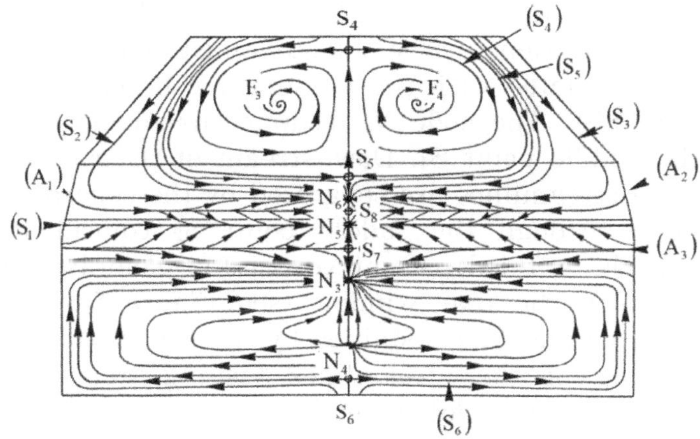

a) First interpretation: one boot vortex

b) Second interpretation: two boot vortices

Figure 5.29. *Surface flow pattern on the rear part of an automobile*

5.4.2. *Separation surfaces*

Figure 5.30(a) shows the separation surface (Σ_1) escaping downstream of the car to constitute the first vortical structure. The streamlines of (Σ_1) originate at the node N_1 that coincides with the saddle point S_1 on the bonnet (see Figure 5.30(a)). These streamlines are split into two families by the separation line (S_1) going through the saddle point S_1 coincident with the node N_5. A second pair of vortices comes from the separation surface (Σ_3) sustained by the separation line (S_4) associated with the passenger cell separation (see Figure 5.30(b)). The node N_3 feeding this surface coincides with the saddle point S_4. Downstream of the car, the two families of streamlines are separated by the separation line (S_2) going through the saddle point S_2 coincident with the node N_6. A third pair of vortices is generated by the separation on the rear window (see Figure 5.30(c)). The separation surface (Σ_3) sustained by the separation line (S_4) is made up of the streamlines issuing from node N_3 coincident with the saddle point S_4.

According to the variants, separation on the base of the boot gives rise to one horseshoe vortex, as shown in Figure 5.30(d) (surface flow pattern of Figure 5.29(a)) or two horseshoe vortices, as shown in Figure 5.30(e) (surface flow pattern of Figure 5.29(b)).

The vortex wake of the car is shown in Figure 5.31, where the separation surfaces of Figure 5.30 have been assembled. The trace in a vertical plane of this eight-vortex system, shown in Figure 5.32, contains:

– the foci F_2 and F_3 corresponding to the separation surface (Σ_3);

– the foci F_4 and F_5 corresponding to (Σ_1);

– the foci F_6 and F_7 corresponding to (Σ_2);

– the foci F_8 and F_9 corresponding to (Σ_4).

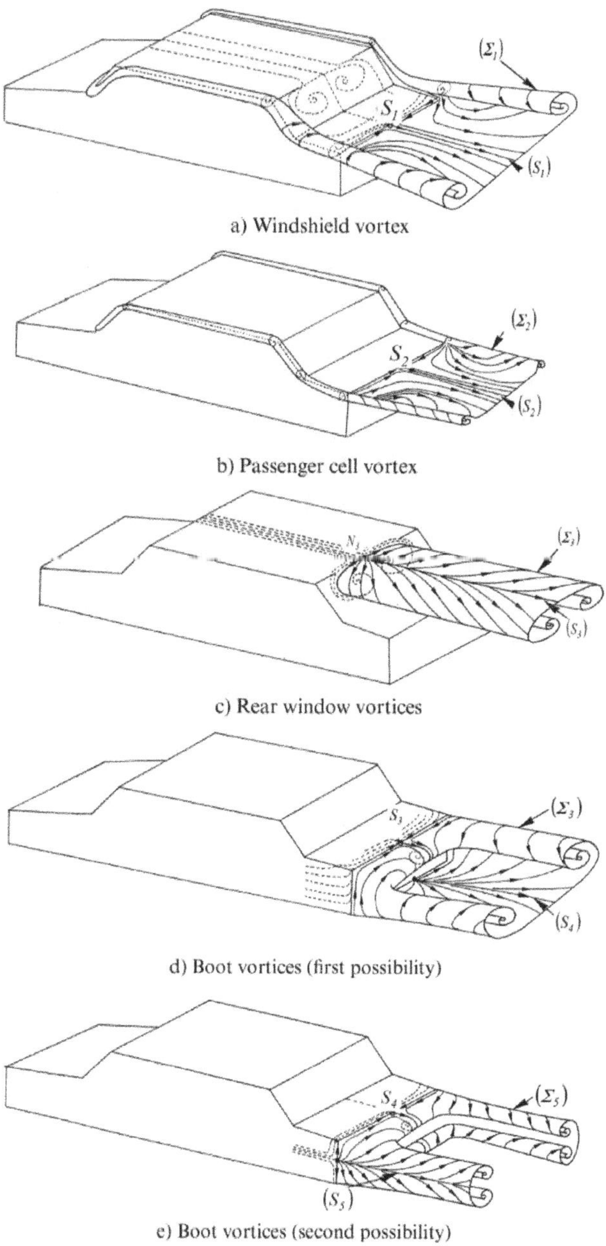

Figure 5.30. *Vortex system emitted by the automobile*

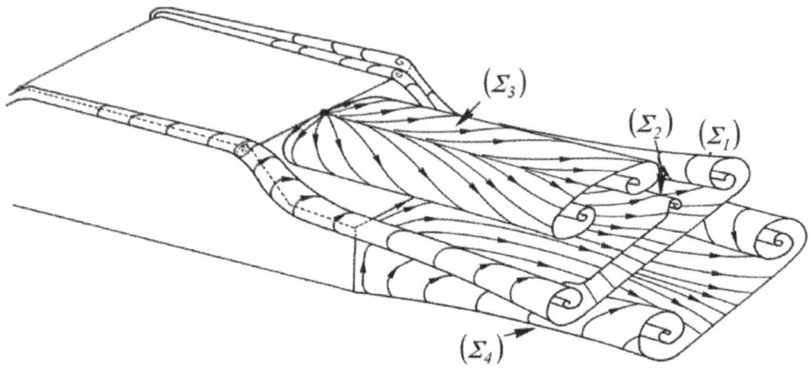

Figure 5.31. *Vortex wake of the automobile*

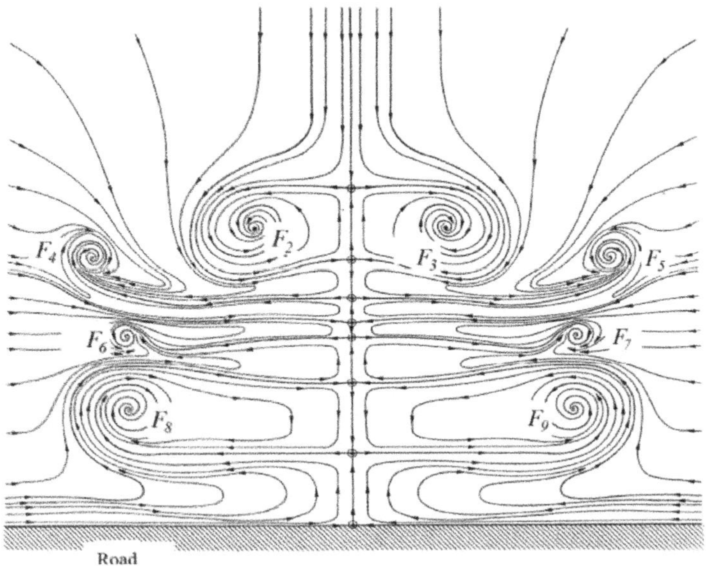

Figure 5.32. *Trace of the vortices in a vertical plane downstream of the automobile*

The other topological components of the projected field can be interpreted without difficulty. As the pseudo-streamlines spiralling into F_8 and F_9 are in contact with the ground, the swirling motion is at the origin of a concentration of dust or water drops emitted by a car moving at high speed on a wet road. By mutual interaction, two vortices close to each other can coalesce in a single structure as for an aircraft wing (see section 4.3.1). This is probably the case for the windshield and passenger cell vortices. Therefore, in reality, the number of foci downstream of the car is less important than shown in Figure 5.32. During this pairing, the intensity – or circulation – of the vortices is invariant so that the vortex system keeps the same total intensity.

To conclude this chapter, Figure 5.33 shows a water tunnel visualization of the flow past on a model of the famous Citroën DS21. The visualization in a symmetry plane shows the attachment nodes on the radiator grille and windshield, a separation on the roof and thc largc vortical region at the vehicle base. We will note the existence of another attachment node below the front part of the car. The photograph gives an instantaneous view of the separated turbulent flow that strongly fluctuates. The previous description applies to a flow that would be steady or to a Reynolds-averaged turbulent flow.

Figure 5.33. *Water tunnel visualization of the flow past on a model of the Citroën DS21*

6

Reconsideration of the Two-Dimensional Separation

The case of the two-dimensional separation (planar and/or axisymmetric) is revisited within the framework of the critical point theory and the topology of three-dimensional separation. In a space with three dimensions, the so-called two-dimensional case appears as highly improbable and inexistent in the reality. Any assumed two-dimensional flow adopts a three-dimensional structure at either a macro- or microscopic scale.

6.1. Some definitions: a reminder

A flow in the three-dimensional space (x, y, z) is called a *planar two-dimensional* flow if its properties are invariant along the coordinate z normal to the plane (x, y). If the velocity vector remains in the plane, the flow is two-dimensional of the first kind; if there is a component along z, the flow is two-dimensional of the second kind. A flow (r, θ, z) is axisymmetric if its properties are independent of the angular coordinate θ. If the velocity vector is contained in the meridian plane, the flow is axisymmetric of the first kind; if there is a component normal to the meridian plane $\theta = Cste$, the flow is axisymmetric of the second kind. This last case corresponds to a projectile stabilized by rotation. Sometimes we consider cylindrical flows whose properties are constant along a direction z that is not normal to the plane (x, y). The component along z can vary with the coordinates (x, y), the condition being that the derivative $\partial/\partial z$ is zero. This kind of flow, corresponding to the swept wing of infinite span, allows important mathematical simplifications while keeping a three-dimensional structure. As the planar two-dimensional and axisymmetric flows of the first kind depend on only two spatial coordinates, they are designated by the generic denomination of two-dimensional flows.

Two-dimensional flows have been extensively studied not only because of the important mathematical simplification resulting from the reduction to two independent space variables, but also because of their easier conceptual perception. However, the existence of two-dimensional flows, as defined above, is unlikely in a three-dimensional world. This is evidenced by experiments that always manifest a three-dimensional organization at either the large or small scale. This difference with respect to the ideal two-dimensional case is interpreted as a parasitic effect where it only reflects reality.

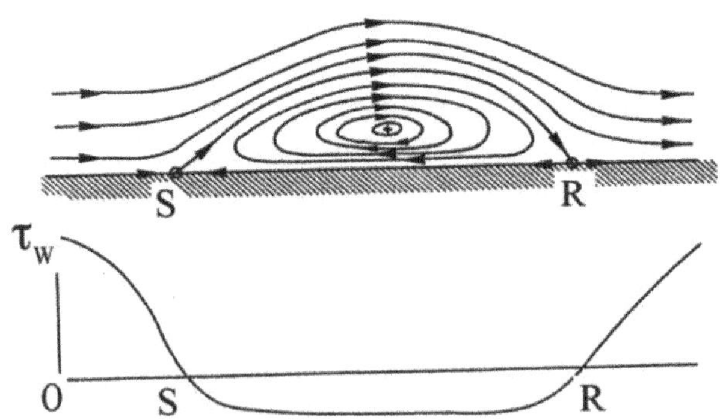

Figure 6.1. *Classical definition of the two-dimensional separation*

6.2. Two-dimensional separation

In a steady two-dimensional flow, separation is defined as the circumstance where at a point S the shear stress at the wall τ_w (considered as a scalar) vanishes. The wall shear stress being positive for an attached flow (by convention), during the separation process τ_w decreases and vanishes at S, τ_w being negative downstream of S, in the separated region. According to circumstances, τ_w goes through a minimum, then rises again and passes through zero at the so-called reattachment point R. Downstream of R, τ_w continues to rise while remaining positive. As shown in Figure 6.1, the streamlines are organized into a bubble inside which they are closed curves, the flow recirculating around a point. A particular streamline (l), connecting S and R, isolates the streamlines in the bubble from those coming from upstream and continuing downstream. The streamline (l) delimits a closed domain inside which the fluid is trapped.

Using the terminology of critical point theory, the bubble is a focus degenerated into a center, the separation and reattachment points being half-saddles and the streamline (l) a separation line. Such a structure can be either two-dimensional, that is invariant with respect to an axis perpendicular to the plane containing the flow, or axisymmetric, that is closed in on itself to form a toroidal vortex.

6.3. Special critical points

To interpret the two-dimensional case in the framework on the three-dimensional concepts, we must consider the critical points associated with special situations in the plane of the parabola $q = p^2 / 4$ (see Figure 6.2).

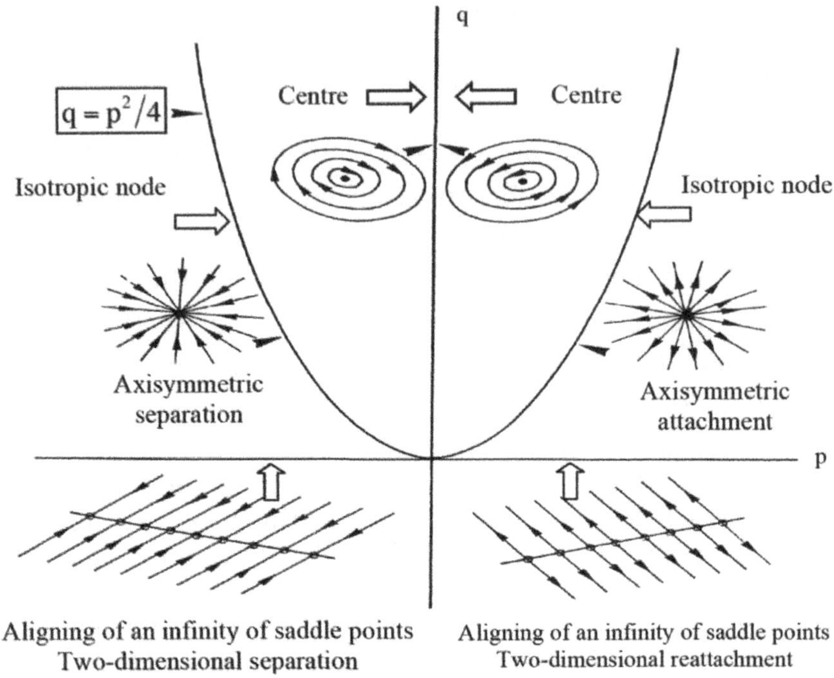

Figure 6.2. *Special situations in the [p, q] plane*

If the representative point is inside the parabola and on the axis $p = 0$; we saw in Chapter 1 that the critical point is a focus degenerated into a center. Then, no skin friction lines go through the critical point, the neighboring lines being closed curved encircling the center. This case corresponds to the classically defined two-dimensional separated bubble. If the point is on the parabola $q = p^2/4$, the critical point is an isotropic node (see section 1.3.2.1), which corresponds to the attachment point on an axisymmetric body at zero incidence or the separation and attachment points of a separated bubble forming inside a flow (the case of the break down of an axisymmetric wake as shown in Figure 6.3).

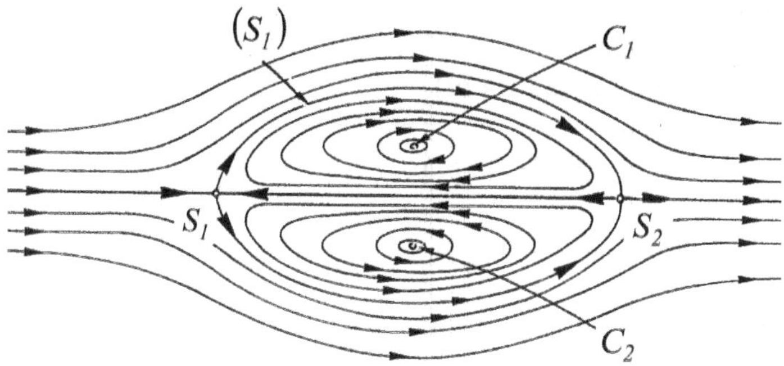

Figure 6.3. *Separation bubble inside an axisymmetric wake*

If the point in the plane $[p, q]$ is on the axis $p = 0$, then the eigenvalues are the roots of equation:

$$S^2 + pS = 0 \rightarrow S(S + p) = 0$$

Hence:

$$S_1 = 0 \quad \text{and} \quad S_2 = -p$$

In this case, the coordinates of the solution curve in the vicinity of critical points are given by:

$$x(t) = \frac{A_1\mu_2}{\lambda_1\mu_2 - \lambda_2\mu_1} - \frac{A_2\mu_1}{\lambda_1\mu_2 - \lambda_2\mu_1}\exp(pt)$$

$$z(t) = -\frac{A_1\lambda_2}{\lambda_1\mu_2 - \lambda_2\mu_1} + \frac{A_2\lambda_1}{\lambda_1\mu_2 - \lambda_2\mu_1}\exp(pt)$$

Putting:

$$B_1 = \frac{A_1}{\lambda_1\mu_2 - \lambda_2\mu_1}, \qquad B_2 = \frac{A_2}{\lambda_1\mu_2 - \lambda_2\mu_1}$$

we can write:

$$x(t) = B_1\mu_2 - B_2\mu_1 \exp(pt)$$

$$z(t) = -B_1\lambda_2 + B_2\lambda_1 \exp(pt)$$

A simple combination gives:

$$\lambda_1 x + \mu_1 z = B_1(\lambda_1\mu_2 - \lambda_2\mu_1)$$

which shows that the skin friction lines are rectilinear curves with the slope:

$$\frac{dz}{dx} = -\frac{\lambda_1}{\mu_1}$$

Thus, parallel trajectories whose slope is independent of B_1 are associated with λ_1 and μ_1. If λ_1 and μ_1 are positive and if p is positive (right part of the axis $q = 0$), when t varies from $+\infty$ to $-\infty$, at infinity the skin friction lines tend to the point of coordinates (see Figure 6.4(a)):

$$x = B_1\mu_2$$

$$z = -B_1\lambda_2$$

For the given eigenvector (λ_2, μ_2), the points such as $x = 0$ and $z = 0$ are on the straight line of slope:

$$\frac{dz}{dx} = -\frac{\lambda_2}{\mu_2}$$

passing through the origin where $B_1 = 0$.

Thus, the straight line of equation

$$\lambda_2 x + \mu_2 z = 0$$

supports an infinity of critical points through which go parallel skin friction lines of slope:

$$\frac{dz}{dx} = -\frac{\lambda_1}{\mu_1}$$

which are travelled from $(x = -\infty, z = +\infty)$ to the critical point if λ_1 and μ_1 are positive. In the case where both λ_1 and μ_1 are negative, the trajectory travels since $(x = +\infty, z = -\infty)$. If $p < 0$, the trajectories are swept from the critical point, as shown in Figures 6.4(b).

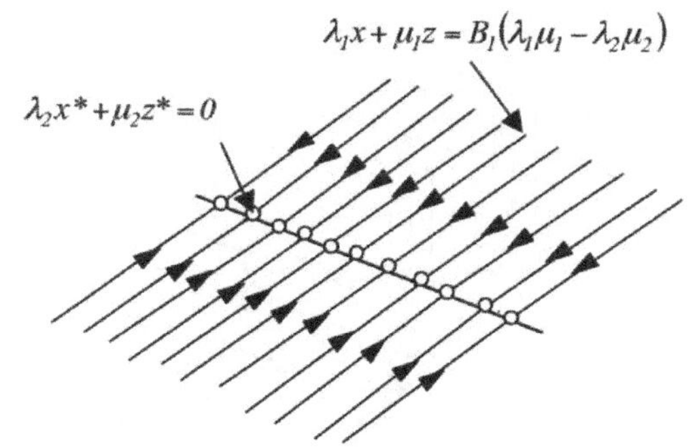

a) Case $p > 0$, convergence motion (separation)

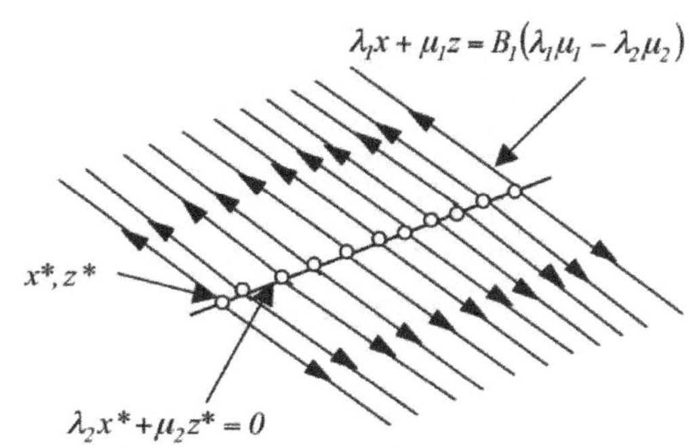

b) Case $p < 0$, divergence motion (attachment)

Figure 6.4. *Particular critical points along an infinite string: case q= 0*

Let us consider the case where the eigenvectors are orthogonal and $\lambda_2 = 0$ (second eigenvector along the axis $z = 0$, which does not restrain generality). Then, the skin fiction lines are perpendicular to the straight line supporting the critical points. Such a circumstance corresponds to a two-dimensional flow where the axis $x = 0$ would be perpendicular to the plane containing the flow. According to the orientation of the trajectories, the critical point is either a re-attachment point (the flow is toward P_0, see Figure 6.5(a)) or a separation point (the flow is away from P_0, see Figure 6.5(b)).

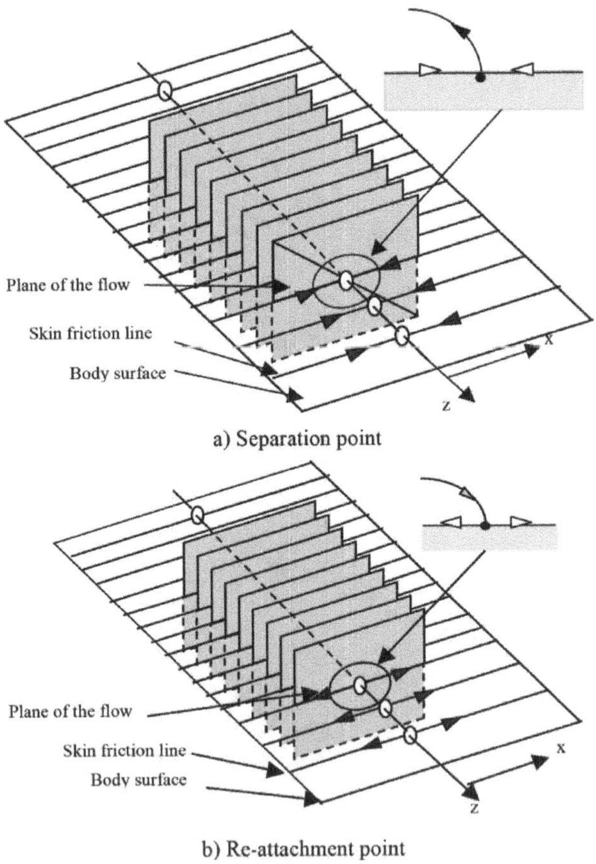

Figure 6.5. *Topological structure of a separating/re-attaching two-dimensional flow*

Thus, a two-dimensional flow is a particular three-dimensional flow in which there exists a straight line perpendicular to the plane of the flow bearing an infinite series of critical points. The two-dimensional case (planar or axisymmetric) is an exceptional circumstance associated with special singularities and organizations. The real world being three-dimensional, we can intuitively perceive that such an arrangement is unrealistic or improbable (instable). The perfectly two-dimensional case supposes that the velocity vector remains in a plane, the streamlines of the separation bubble being perfectly closed curves. However, if the velocity has a component (even very small) outside the plane, the streamlines are no longer closed curves, and the fluid escapes laterally.

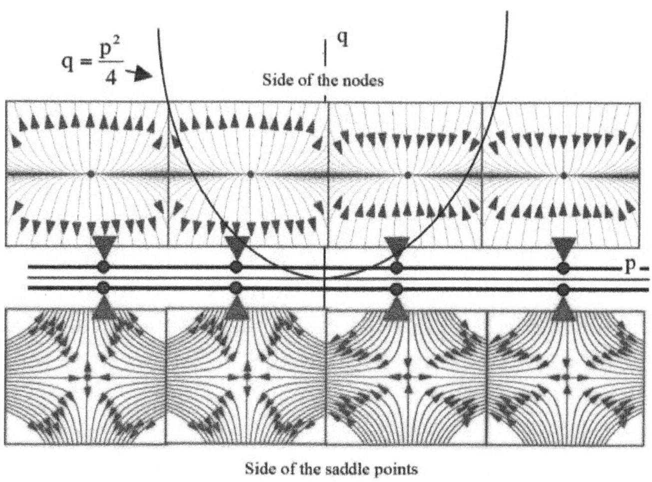

Figure 6.6. *Limit situation in the plane of the parabola $q = p^2/4$. Switching from the node side to the saddle point side. For a color version of this figure see www.iste.co.uk/delery/3Dflow.zip*

In reality, the fluid organizes itself by adopting a structure where the critical points are of finite number and distinct, which gives to the flow a certain freedom for a lateral motion. The critical points, the image of which are on the axis $q = 0$ in Figure 1.3, have a double nature: they belong both to the region of the nodes and that of the saddle points because their image is on the boundary between these two regions. Because two nodes or two saddle points of the same nature (separation or attachment) cannot be connected on the same skin friction line, the flow on the separation (or attachment) line alternatively switches on the nodes or saddle points side (see Figure 6.6). The separation (attachment) line presents a string of nodes and saddle points as shown in Figure 6.7. According to the orientation of skin friction lines, the surface flow pattern is associated with either a separation or an attachment process. Such a structure where two nodes are separated by a saddle point is topologically consistent.

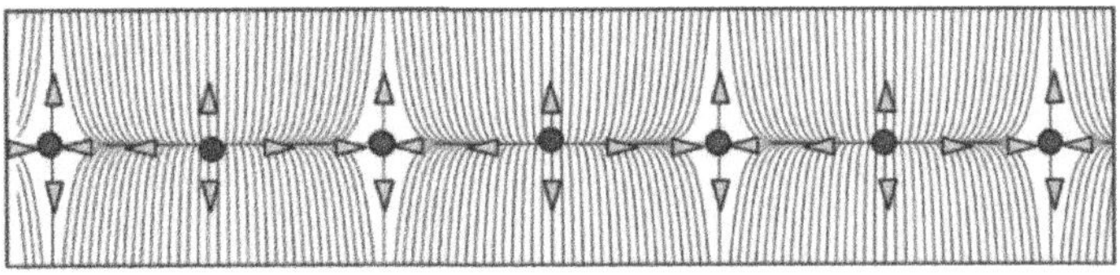

Figure 6.7. *Vicinity of axis $q = 0$ or when nodes and saddle points meet*

Such a structure is observed in axisymmetrical flows where the surface flow visualizations reveal an alternation of nodes and saddle points both at separation and re-attachment as shown in Figure 6.8 relative to a supersonic re-attachment on a flare.

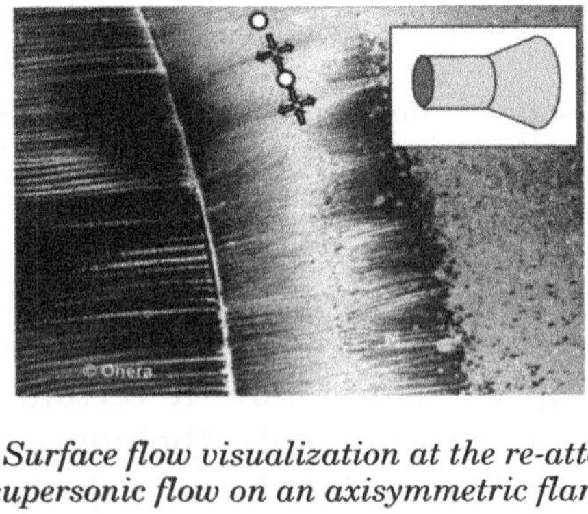

Figure 6.8. *Surface flow visualization at the re-attachment of a supersonic flow on an axisymmetric flare*

6.4. Three-dimensional structure of a two-dimensional separated flow

The surface flow pattern adopts a fine structure where the separation line, or attachment line, bears a succession of a finite number of node/saddle point combinations as shown in Figure 6.9. The nodes on (A) are attachment nodes feeding partly the downstream flow and partly the upstream flow. The same periodic structure exists on the separation line (S), the pattern being organized in a succession of saddle points and separation nodes. If the separated flow between (S) and (A) is "closed", the skin friction lines issued from the nodes of (A) terminate in the nodes of (S), as shown in Figure 6.10. As (A) and (S) sustain the same number of nodes and saddle points, equation [3.2] is satisfied if the nodes at upstream and downstream infinity are accounted for. The surface flow pattern is divided into independent cells separated by separation lines going through the saddle points. Then, the separation line (S) sustains the separation surface (Σ) ending on the attachment line (A), as shown in Figure 6.11. The streamlines of (Σ) are also organized in cells and the saddle points of (Σ) coincide with the nodes of (S) and (A) (and reciprocally). On (Σ), the saddle points and the nodes are linked by separation lines such as (S_1) and (A_1) (see Figure 6.12). These lines are in the planes (P_1) and (P_2) containing the separation lines (A_1) and (S_1) of the surface pattern. The volume of separated fluid is cut up into slices that do not communicate.

The recirculation bubble bounded by (Σ) is constituted by a vortex whose sections by meridian planes of the axisymmetric body are centers (these centers are, in fact, relative to the pseudo-streamlines projected in the considered plane). This means that the vortex streamlines wind into spirals on toroidal surfaces as shown in Figure 6.12. Like a spring, the spires move apart from a plane like (P_1) and contract when they approach (P_2). If we imagine that the saddle points and nodes lying on (S) and (A) tend to be coincident, then each sector of the vortex flattens out so that the flow recovers the organization of the purely two-dimensional case made up of a piling of streamlines organized around centers.

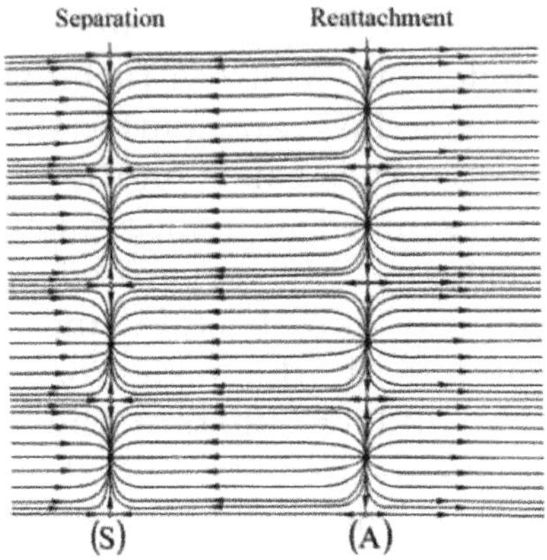

Figure 6.9. *Three-dimensional structure of a two-dimensional separated flow*

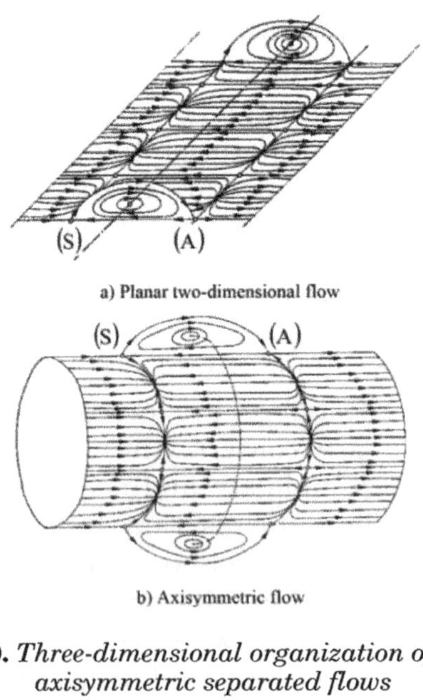

Figure 6.10. *Three-dimensional organization of planar and axisymmetric separated flows*

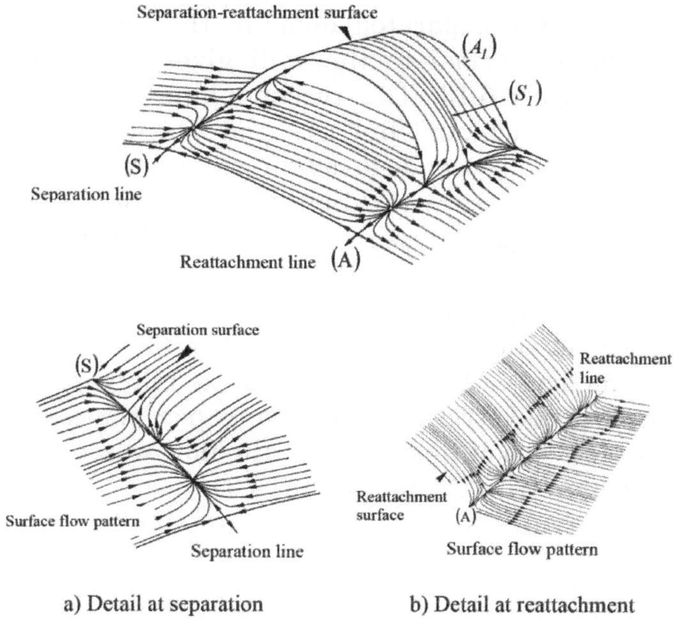

Figure 6.11. *Three-dimensional structure of a two-dimensional separated flow. Separation / reattachment surface*

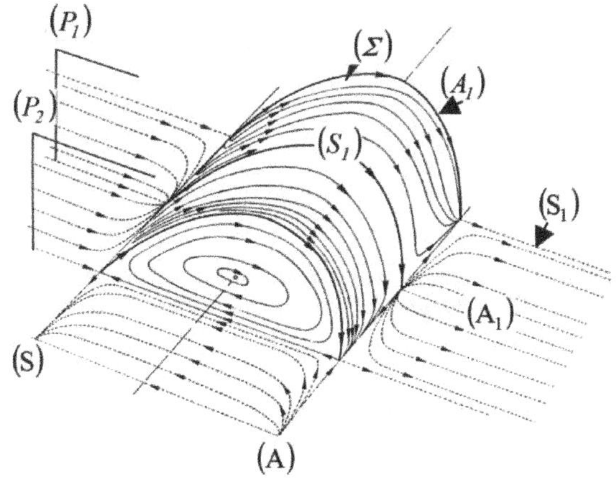

Figure 6.12. *Three-dimensional structure of a two-dimensional separated flow. Organization of the recirculation bubble. For a color version of this figure see www.iste.co.uk/delery/3Dflow.zip*

Outside of the separated domain, the projection of the flow in planes normal to the wall gives the organization shown in Figure 6.13. The chosen planes are (P_3), which contains the attachment line (A), and (P_4), which is located downstream. The pseudo-streamlines are organized in a succession of centers with alternate sense or rotation; this structure is compatible with the surface flow pattern. The motion having a streamwise component, the streamlines are organized in cylindrical streamwise vortices, parallel between them and rotating in an opposite sense (contra-rotating vortices). We must imagine a piling of a certain number of these vortices in order to respect the continuity of the field. Analysis of this aspect of the flow would take us too far (connection with Görtler' vortices?).

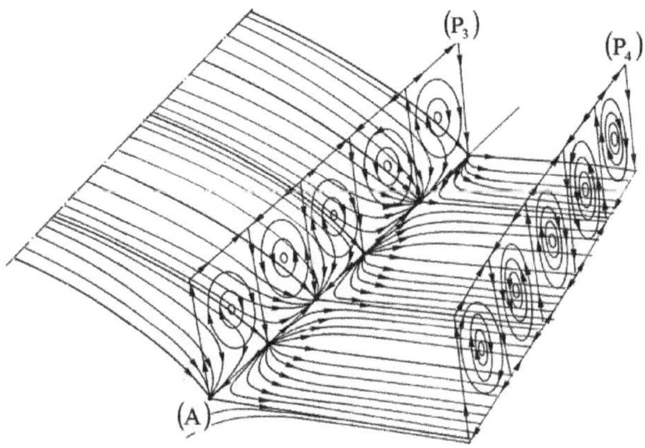

Figure 6.13. *Three-dimensional structure of a two-dimensional separated flow. Organization of the outer flow*

The structures depicted above are rarely observed in nominally planar two-dimensional flows. As a vortex must either close on itself or terminate on a material wall or a free surface, the vortex produced by separation should extend indefinitely in the spanwise direction, that is it closes on itself at infinity. As all the real devices have a finite span, the observed surface flow pattern has, in fact, the structure shown in Figure 6.14. The separation line going through the saddle point S_1 located in the middle of the test section spirals into vortices that are the traces on the surface of tornado-like vortices forming near the sidewalls. The fluid that cannot remain trapped in a closed bubble escapes by these vortices, the re-attachment line being the trace of an attachment surface distinct from the separation surface. If the test section span is large compared to a representative dimension of the flow (for instance, the thickness of the incoming boundary layer), the separation and re-attachment lines are nearly rectilinear in the central part of the installation, and we say that the flow is "practically two-dimensional". In reality, since a lateral evacuation of the flow must exist, we have to be prudent when utilizing experimental results obtained in such conditions to validate two-dimensional theoretical models. A more complete description of the real flow topology would require us to take into consideration the phenomena on the sidewalls of the test setup.

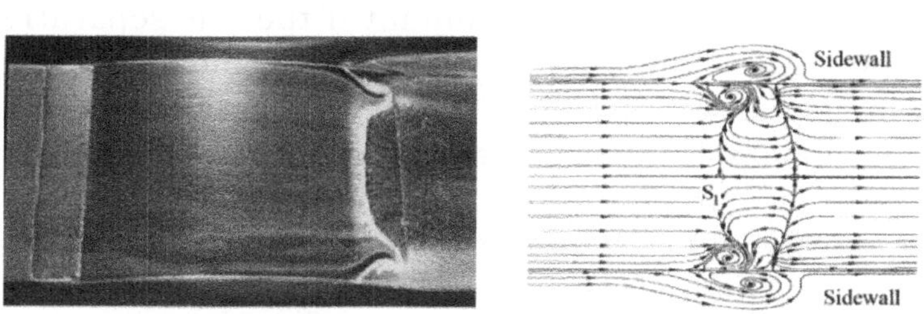

a) Surface flow visualisation b) Topological interpretation

Figure 6.14. *Separation in a two-dimensional transonic channel (courtesy of IMP-Gdansk)*

Such a flow organization is frequently considered as representative of three-dimensional effects to mean that it is a perturbation of a flow that we would wish to be two-dimensional. In this perspective, these effects can be termed *macroscopic* three-dimensional effects because they affect a large part of the whole flow field. However, in axisymmetric separated flows, where side effects are absent, the cellular organization is observed. Because of the lack of a possibility of adjustment allowing it to laterally escape, the flow adopts such an organization that appears as the most probable if axisymmetry must be respected. In this case, the three-dimensional effects will be termed *microscopic* because they affect the flow field locally. Then, the two-dimensional approximation may faithfully represent reality.

The two-dimensional assumption is mainly a mathematical convenience because two-dimensional devices are very rare. There always exists a side limitation making the flow three-dimensional. The so-called two-dimensional supersonic air intakes have, in reality, a rectangular cross-section so that phenomena on the sidewalls, especially in the corners, affect the flow inside the air intake. This influence can be weak if there is no separation so that the two-dimensional approximation can be safely applied (case of the compression ramps of a supersonic air intake if the shock waves are not strong enough to induce separation). However, this influence becomes predominant if there is separation so that a three-dimensional model becomes mandatory. Axisymmetric devices, such as axisymmetric air intakes, propulsive nozzles, missile afterbodies (if control surfaces are omitted), and circular pipes, are more frequent. Then, the two-dimensional (axisymmetric) approximation is justified if the incidence angle is zero or very small.

6.5. Axisymmetric afterbody

We consider an axisymmetric afterbody, without propulsive jet, set at zero incidence in a uniform flow. In the ideal axisymmetric case, the flow is the same in each meridian plane. As shown in Figure 6.15(a), the field in a meridian plane contains two centers C_1 and C_2, three half-saddle points S_1, S_2 and S_3 on the wall, and a saddle point S_4

in the flow field, downstream of the base, often called a re-attachment point. The separation lines (S_1) and (S_2) issued from S_1 and S_2 meet at the saddle point S_4 enclosing a fluid trapped in two recirculation regions constituted by the centers C_1 and C_2. The base flow is organized as a toroidal vortex whose axis supports an infinity of centers.

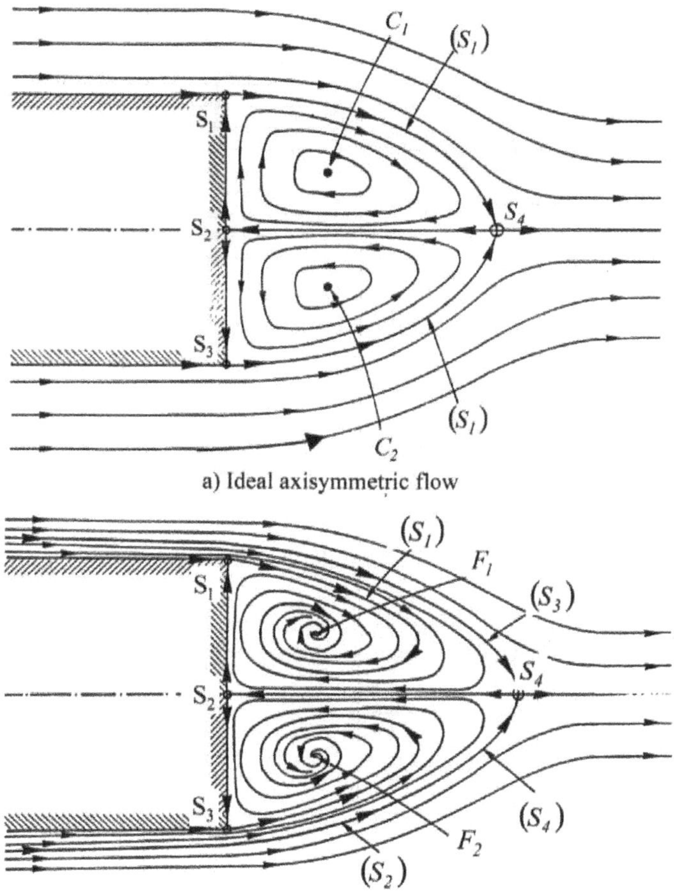

a) Ideal axisymmetric flow

b) Real flow

Figure 6.15. *Flow past a perfectly axisymmetric afterbody. Meridian field*

An isotropic attachment node N_1 is located on the base center. The skin friction lines emanating from this node flow toward the base shoulder that supports an infinity of detachment nodes through which the skin friction lines coming from upstream and N_1 are evacuated (see Figure 6.16(a)). A separation line (S_1) along the base shoulder sustains the separation surface closing at the saddle point S_4 of Figure 6.15(a) (see Figure 6.17(a)).

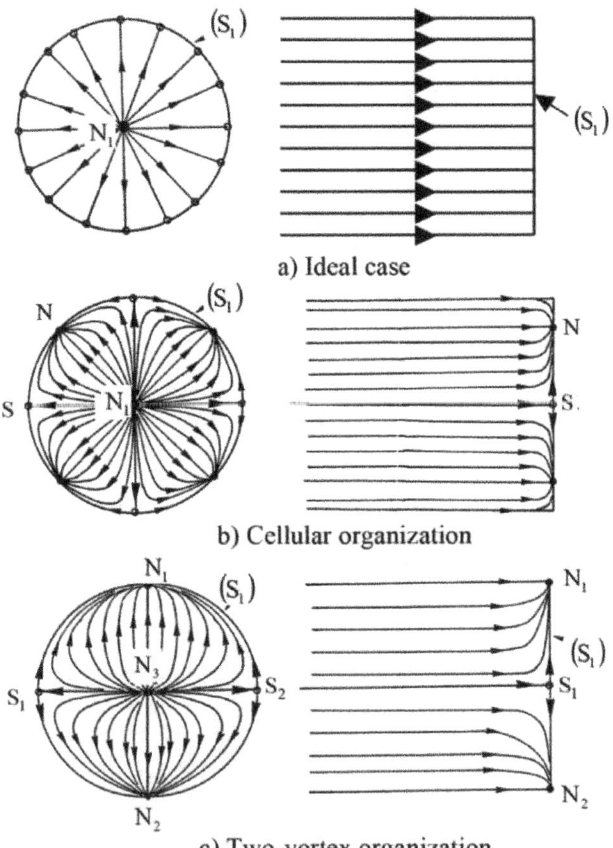

Figure 6.16. *Real axisymmetric flow with cellular organization. Surface flow pattern*

The previous situation being improbable, in reality a three-dimensional organization establishes itself. A first possibility is the structure with periodic cells shown in Figure 6.16(b), which preserves a quasi-axisymmetric organization. The separation surface (Σ) is now constituted of streamlines adopting a periodical arrangement (see Figure 6.17(b)). Coming from the nodes located on the base shoulder, they terminate in the attachment node N_1 on the centerline. Figure 6.16(b) shows the surface flow pattern on the base: the skin friction streamlines originating at the node N_1 located on the center flow in the direction of the separation nodes on the base shoulder. Separation lines, going through the saddle points of (S_1), split these lines into sectors whose number depends on the general flow circumstances. The base flow is organized in a toroidal vortex, the structure of which is identical to that depicted in section 3.3. Sections of this vortex by meridian planes are centers, the fluid in the base region being isolated from the rest of the flow by the separation surface (Σ).

This organization being rather special, the flow (under the effect of a small lack of axisymmetry) probably adopts the structure shown in Figure 6.17(c) where the centers are replaced by two foci F_1 and F_2 located in the vertical meridian plane (see Figure 6.15(b)). We again find the saddle points S_1, S_2, S_3 and S_4 as well as the separation lines (S_1), (S_2) and (S_3). Now, the separation lines going through the saddle point S_4 are distinct from (S_1) and (S_2), the base flow being "open". As shown in Figure 6.16(c), the skin friction lines coming from the node at "upstream infinity" are split into two families by the separation lines (S_1) and (S_2) going through S_1 and S_2; these families flow to the separation nodes N_1 and N_2. An attachment node N_3 on the base center is the source of the skin friction lines terminating in N_1 and N_2, the splitting being made by the separation line going through S_1 and S_2. We can check that relation [3.2] is satisfied (taking into account a node at upstream infinity). The separation line (S_1) along the base shoulder sustains the separation surface (Σ) that wraps into two horseshoe vortices as shown in Figure 6.17(c). The traces of these vortices in a vertical meridian plane are the foci F_1 and F_2. The streamlines of (Σ) originate at the attachment nodes N_1 and N_2 in coincidence with the saddle points S_1 and S_2. These lines are split into two families, one spiraling around the upper vortex and the other spiraling around the lower vortex. Figure 6.18 shows the organization of the pseudo-streamlines of the field projected in a plane normal to the axisymmetry axis.

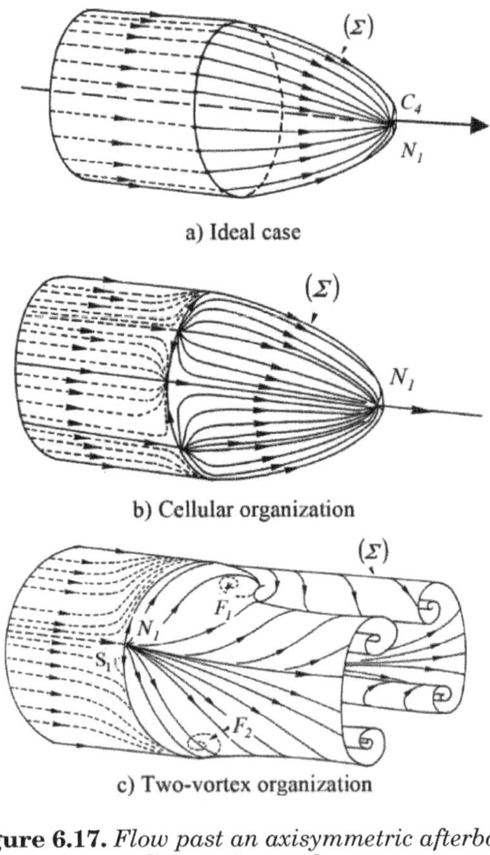

Figure 6.17. *Flow past an axisymmetric afterbody. Separation surface*

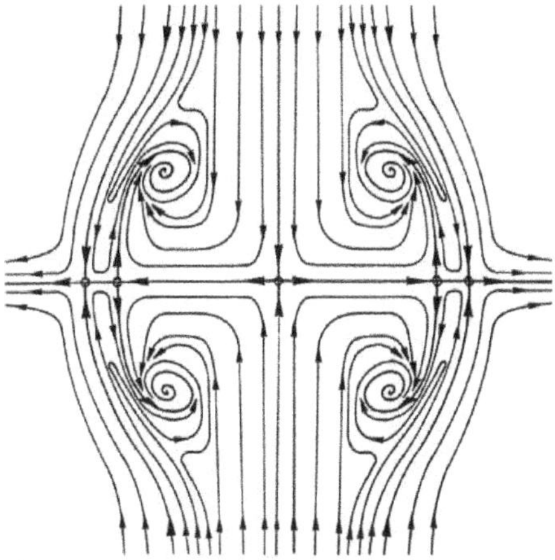

Figure 6.18. *Two-vortex base flow. Field projected in a downstream plane*

Such a configuration having a high degree of symmetry must be instable so that a slight defect of axisymmetry can make the configuration switch to the structure with a unique horseshoe vortex shown in Figure 5.19.

7

Concluding Remarks

The conclusion stresses the practical importance of separation as a cause of performance degradation and the source of nuisances. The concepts deduced from the critical point theory constitute a grammar allowing a rational description of separated flows met in most devices using fluids. The topological consideration could be extended without difficulty of principle to unsteady flows.

The aim of this book was to present the theoretical notions allowing a rational interpretation of three-dimensional velocity fields, especially when they contain separated regions. The basis of the interpretation is the critical point theory that allows for introducing special points such as nodes, saddle points and foci, observed in the skin-friction line pattern, and linked to the separation process. Then, we defined separation lines and separation surfaces leading to the concepts of attachment, separation (or detachment) and vortical structures. The set constitutes a logical grammar allowing a precise definition of the separation phenomenon as well as a consistent topological description of three-dimensional flows, which are, assumed here, steady.

These principles are applied to the construction of typical separated flows forming past wings, fuselages, in front of obstacles, and around protuberances. We also consider the flow past a non-propelled afterbody set at incidence and an automobile. The two-dimensional case is considered in the light of the general three-dimensional concepts.

Separation, with its many inconveniences (increased drag, reduced efficiency, occurrence of unsteadiness and vibrations, noise generation, increased local thermal loads, etc.) remains the major concern of aerodynamicists and, more greatly, of the designers of devices operating fluids. Only the fluids termed perfect, that is non-viscous, ignore the separation phenomenon (except in very special circumstances). However, the perfect fluid concept is a limit situation non-existent in reality, the dropping of viscosity being a mathematical convenience. In reality, the occurrence of separation is due to the existence of viscosity that is always present, even in nearly perfect nominal conditions, such as the trailing edge of an airfoil at low incidence. On the other hand, viscosity, if a good thing, is because it is at the origin of lift and causes the dissipation into heat of the flow momentum which otherwise would cause trouble.

Not considering the separation node at the trailing edge; it is rare to find a situation where separation, as defined here, does not exist. In fact, a saddle point in the surface flow pattern appears at the rapid variation of a contour (a base shoulder, for example) or under the action of an adverse pressure field, even of modest intensity. However, by concluding that all flows are separated, there is the risk of being in a situation where the present analysis loses any operational value because wings, diffusers, and air intakes working correctly evidently exist. We can introduce a difference between the concept of *separation*, resulting from the occurrence of a saddle point, which is a common feature, and that of *detachment* when separation takes a catastrophic character due to the brutal surging of vortices into the flow field. The engineer is only interested in the second situation, which distinguishes itself from the general case by the intensity of the phenomena. For this reason, the critical point theory and the concepts of lines of separation and separation surfaces keep their interest as an interpretation tool and, consequently, help conception.

The critical point theory has an unjustified reputation of complexity. In fact, it is based on elementary mathematical developments having the advantage of leading to results that can be directly observed in surface flow patterns or velocity fields. If the construction of some flow field is considered complex, it is because the perception of three-dimensional objects is difficult. However, this theory allows a powerful generalization of the notion of separation and allows us to look beneath the new aspect phenomena whose interpretation with the usual concepts is not convincing and sometimes acrobatic.

The critical point theory shows that the difficulty is, in fact, with the notion of two-dimensional flow whose realization in the physical world is highly improbable. Hence, the danger to try to explain situations with concepts applying to flows that, in fact, do not exist and to force nature by inventing experiments that are alleged two-dimensional. The huge number of experiments on two-dimensional profiles, ramps, upstream or downstream facing steps has been useful to interpret the physical phenomena and define the correlation laws. However, they can be safely exploited only by running three-dimensional Navier–Stokes codes where the test conditions are represented as faithfully as possible. Only axisymmetric models produce nearly two-dimensional flows if the angles of incidence and yaw are perfectly set to zero. However, as seen in Chapter 6, even in this case, the three-dimensional character of nature manifests itself as microstructures, the influence of which does not substantially affect the overall flow axisymmetry.

We may object that if two-dimensional flows do not exist, steady flows do not exist either. Cancelling the time derivatives is a mathematical convenience that not only prevents the correct description of reality but is also an obstacle to the correct understanding of some phenomena. For example, shock phenomena can only be understood with a good familiarity with the propagation process. With the rapid progress in computation methods, the solution of the full Navier–Stokes equations has become usual to describe turbulence or predict large-scale unsteady phenomena such as stalling, and buffeting of buzz. The critical point theory applies to a vector field that is an average view in the usual approach. However, this field can be an instantaneous field in a temporal vision of reality, such as the picture given by either computation or measurements (with particle image velocimetry (PIV)). To keep the size of this book reasonable, we restricted ourselves to steady flow fields, although there is no difficulty in principle to consider unsteady frozen pictures, such as the Kàrmàn alley downstream of a cylinder (see Figure 7.1).

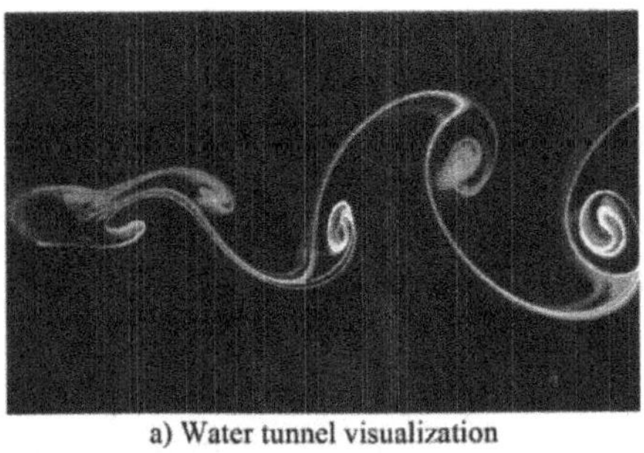

a) Water tunnel visualization

b) Topology of the instantaneous field

Figure 7.1. *Vortex alley downstream of a cylinder*

Bibliography

[BAK 89] BAKKER P.G., WINKEL M.E.M., "On the topology of three-dimensional separated flow structures and local solutions of the Navier-Stokes equations", *Proceedings of the IUTAM Symposium on Topological Fluid Mechanics*, Cambridge, 13–18 August 1989.

[CHA 86a] CHANETZ B., Contribution à l'étude du décollement tridimensionnel en écoulement turbulent, PhD Thesis, University Claude Bernard of Lyon, September 1986, and Onera Technical Note 1988-6.

[CHA 86b] CHAPMAN G.T., Topological classification of flow separation on three-dimensional bodies, AIAA Paper 86-0485, January 1986.

[DAL 83] DALLMANN U., Topological structures of three-dimensional flow separation, Report DFVLRE – IB 221-872, 1983.

[DÉL 93] DÉLERY J., "Physics of vortical flows", *Journal of Aircraft*, vol. 29, no. 5, pp. 856–876, 1993.

[DÉL 94] DÉLERY J., "Aspect of vortex breakdown", *Progress in Aerospace Sciences*, vol. 30, no. 5, pp. 1–59, 1994.

[DÉL 95] DÉLERY J., "Fundamental phenomena related to 3D afterbodies", *Aerodynamics of 3D Aircraft Afterbodies*, AGARD AR 318, pp. 7–24, 1995.

[DÉL 01] DÉLERY J., "Robert Legendre and Henri Werlé: toward the elucidation of three-dimensional separation", *Annual Review of Fluid Mechanics*, vol. 33, pp. 129–154, 2001.

[DÉL 10] DÉLERY J. *Handbook of Compressible Aerodynamics*, ISTE Ltd, London and John Wiley & Sons, New York, 2010.

[DOE 89] DOERFFER P., DALLMANN U., "Reynolds number effect on separation structures at normal shock wave/turbulent boundary layer interaction", *AIAA Journal*, vol. 27, no. 9, pp. 1206–1212, 1989.

[EAS 68] EAST L.F., HOXEY R.P., Boundary layer effects in an idealized wing-body junction at low speed, RAE TR 68161, 1968.

[GIN 58] GINOUX J., Experimental evidence of three-dimensional perturbations in the reattachment of a two-dimensional laminar boundary layer, Von Karman Institute Report TN-1, 1958.

[HOR 84] HORNUNG H., PERRY A.E., "Some aspects of three-dimensional separation. Part 1: streamsurface bifurcations", *Zeitschrift für Flugwiss. Weltraumforsch*, vol. 8, pp. 77–87, 1984.

[HUN 78] HUNT J.C.R., ABELL C.J., PETERKA J.A., WOO H., "Kinematical studies of the flows around free or surface-mounted obstacles; applying topology to flow visualization", *Journal of Fluid Mechanics*, vol. 86, part 1, pp. 179–200, 1978.

[LEG 52] LEGENDRE R., "Ecoulement au voisinage de la pointe avant d'une aile à forte flèche aux incidences moyennes", *La Recherche Aéronautique*, no. 30, pp. 3–8, 1952.

[LEG 56] LEGENDRE R., "Séparation de l'écoulement laminaire tridimensionnel", *La Recherche Aéronautique*, no. 54, November–December 1956.

[LEG 65] LEGENDRE R., "Lignes de courant d'un écoulement continu", *La Recherche Aérospatiale,* no. 105, March 1965.

[LEG 66] LEGENDRE R., "Vortex sheet rolling up along leading edges of delta wings", *Progress in Aerospace Sciences*, vol. 7, pp. 7–33, 1966.

[LEG 77] LEGENDRE R., "Lignes de courant d'un écoulement permanent. Décollement et séparation", *La Recherche Aérospatiale*, no. 1977-6, November–December 1977.

[LIG 63] LIGHTHILL J., "Attachment and separation in three-dimensional flows", in ROSENHEAD L. (ed.), *Laminar Boundary Layer Theory*, Section II, Chapters 2–6, Oxford University Press, New York, pp. 73–82, 1963.

[PEA 80] PEAKE D.J., TOBAK M., "Three-dimensional interactions and vortical flows with emphasis on high speed", *AGARDograph*, no. 252, March 1980.

[PER 74] PERRY A.E., FAIRLIE B.D., "Critical points in flow patterns", *Advance in Geophysics*, vol. B-19, pp. 200–315, 1974.

[PER 84] PERRY A.E., HORNUNG H., "Some aspects of three-dimensional separation. Part II: vortex skeleton", *Zeitschrift für Flugwiss. Weltraumforsch*, vol. 8, pp. 155–160, 1984.

[POI 91] POINCARÉ H., "Les points singuliers des équations différentielles", *Comptes-Rendus de l'Académie des Science*, 13-2-1891, 1891 and *Œuvres Complètes*, vol. 1, 1891.

[SED 58] SEDNEY R., KITCHENS C.W., "Separation ahead of protuberances in supersonic turbulent boundary layers", *AIAA Journal*, vol. 15, pp. 546–552, 1958.

[TOB] TOBAK M., PEAKE D.J., "Topology of three-dimensional separated flow", *Annual Review of Fluid Mechanics*, vol. 14, pp. 61–85, 1982.

[VAN 82] VAN DYKE M., *An Album of Fluid Motion*, Parabolic, Stanford, CA, 1982.

[WAN 75] WANG K.C., "Boundary layer over a blunt body at low incidence with circumferential reversed flow", *Journal of Fluid Mechanics*, vol. 72, part 1, pp. 49–65, 1975.

[WER 75] WERLÉ H., Ecoulements décollés. Etudes phénoménologique à partir de visualisations hydrodynamiques, Onera, TP 1975-14, 1975.

[WER 77a] WERLÉ H., Le tunnel hydrodynamique au service de la recherche aérospatiale, Onera, Publication no. 156, 1977.

[WER 77b] WERLÉ H., "Bulbes de recirculation au sein d'un sillage", *La Recherche Aérospatiale,* no. 1, January–February 1977.

[YAT 92] YATES L.A., CHAPMAN G.T., "Streamlines, vorticity lines and vortices around three-dimensional bodies", *AIAA Journal*, vol. 30, no. 7, pp. 1819–1826, July 1992.

List of Symbols

Basic notations

C_L:	lift coefficient
p:	trace of the Jacobian matrix
p:	static pressure
p:	complexity of a surface
q:	discriminant of an eigenvalue equation
r, θ, z:	cylindrical coordinates
S_1, S_2:	eigenvalues
t:	integration parameter
u, v, w:	velocity Cartesian components
$\vec{V}$:	velocity vector
V_1, V_2:	Cartesian velocity vector components
V_r, V_θ, V_z:	velocity components in cylindrical coordinates
x, z:	Cartesian coordinates on the body surface
y:	coordinate normal to the surface
β:	boundary layer distortion angle

Γ: circulation

λ, μ: eigenvectors components

ρ: density

τ_w: skin friction vector at the wall

τ_x, τ_z: skin friction vector components

$\vec{\Omega}$: vorticity vector

$\vec{\Omega}_w$: vorticity vector at the wall

$\omega_x, \omega_y, \omega_z$: vorticity vector components

Topological features

The adopted convention is to designate features belonging to the surface flow by roman characters and features belonging to the outer flow by italic characters. Outer flow special surfaces are designated by Greek letters, such as (Σ) for a separation surface.

(A), (*A*): attachment line

(*F*): vortex centerline

F, *F*: focus

(*f*): skin friction line

(*l*): streamline

N, *N*: node

(P): projection plane

S, *S*: saddle point

(S), (*S*): separation line (separatrix)

(Σ): separation (attachment) surface

Index

Printed and bound by CPI Group (UK) Ltd, Croydon, CR0 4YY

13/09/2026

14961075-0002